Home in the Howling Wilderness

Home in the Howling Wilderness

SETTLERS AND THE ENVIRONMENT IN SOUTHERN NEW ZEALAND

PETER HOLLAND

AUCKLAND
UNIVERSITY
PRESS

First published 2013

Auckland University Press
University of Auckland
Private Bag 92019
Auckland 1142
New Zealand
www.press.auckland.ac.nz

© Peter Holland, 2013

ISBN 978 1 86940 739 1

Publication is kindly assisted by the

National Library of New Zealand Cataloguing-in-Publication Data
Holland, Peter, 1939-
Home in the howling wilderness : settlers and the environment in
Southern New Zealand / Peter Holland.
Includes bibliographical references and index.
ISBN 978-1-86940-739-1
1. Agriculture—New Zealand—South Island—History—19th century.
2. Human ecology—New Zealand—South Island. 3. Nature—Effect
of human beings on—New Zealand—South Island. 4. Colonists—
New Zealand—South Island. 5. South Island (N.Z.)—Environmental
conditions. I. Title.
630.9937—dc 23

Front cover: 'The storm – Lake Wanaka – NZ', c. 1935. George Chance
photograph, Hocken Collections, Uare Taoka o Hākena, University of Otago.
Reproduced courtesy of the George Chance estate
Back cover: Two four-horse ploughs, Waimate, south Canterbury.
Waimate Historical Society and Museum, 2002-1026-00052

Cover design: Kalee Jackson

Printed in China by Everbest Printing Co. Ltd

CONTENTS

He found him in a desert land, and in the waste howling wilderness; he led him about, he instructed him, he kept him as the apple of his eye.

– DEUTERONOMY 32:10

ACKNOWLEDGEMENTS

I first thought about this book 30 years ago, and often talked about it with my former colleague Professor Sherry Olson of McGill University in Montréal. That it reached publication after such a long voyage owes much to her continuing support and quiet insistence. Several other people were also generous with their time and knowledge during the early years of this project. An uncle, the late Charles Wall, told me about life in rural south Canterbury during the 1930s, forties and fifties, and answered my questions about sheep and cattle, fences and hedges. More recently, Robert Holland has been my source of information about sheep, pastures and the daily life of a practical farmer.

For the past twenty years I have visited museums and archives in search of documentary material about farmers and the land. Five hosted me for periods of days to weeks at a time, gave me access to their archives and a place to work, and answered my questions about nineteenth-century farms and stations. Staff of the Hocken Collections at the University of Otago helped me trace manuscript diaries and letters, permitted me to quote from them, and provided a congenial research environment that added to the pleasures of the hunt. They also made available digital copies of photographs from their collection and granted permission to publish them. In Oamaru, staff of the North Otago Museum provided access to their holdings of local historical material and allowed me to draw on it for this research. But it was across the Waitaki River, in my home town of Waimate, where I first became almost obsessive about archival research. Late one Friday afternoon, just as I was about to wrap up a productive week in the library of the Waimate Historical Society and Museum and drive back to Dunedin, one volunteer staff member said, 'There's something in the attic that might interest you.' He was right. There, under a sheet of canvas, was a metre-high pile of ledger books from the now defunct south Canterbury firm Manchester and Goldsmith, and the entries related to Te Waimate Station in the 1870s and 1880s. Thank you, Murray! For the next two years, whenever I could take several days away from my office in Dunedin, I spent many hours transcribing entries from them. I have still to make full use of that material, but that's my next job. Museum staff were generous with their time and helpful in many other ways as well, including granting me permission to reproduce photographs from their collection. The extensive holdings of newspapers, original documents and photographs in the South Canterbury Museum, Timaru, also proved invaluable, and I am grateful to the staff for making them available and giving me permission to reproduce several photographs from their archive. Members of staff of the Canterbury Museum were of great assistance during the early years of this project, and since then have provided copies of diaries in their collection and granted permission for me to quote from them.

Part of the research described here was undertaken with support from a Marsden Grant (principal investigators Professors Tom Brooking and Eric Pawson). I have benefited from frequent discussions with Charles Forsyth, Frank Leckie and Gordon Parsonson, as well as with Drs George Davis and Guil Figgins. The Department of Geography provided a grant for publication costs, and generously made available a computer as well as work and study space; Christine Bradshaw, Nigel McDonald and David McDowall came to my rescue on numerous occasions when my ignorance of software looked set to scuttle the project; Tracy Connolly drafted the figures from my rough sketches; Philippa Dixon and Vaughan Wood provided copies of several diaries and assisted in locating other unpublished material; Dr Jim Williams made available copies of archival material and read a draft of Chapter 1; and Dr John Morrissey commented on drafts of other chapters, identified areas where more information was needed, and encouraged me to finish.

Dr Ginny Sullivan was a gracious and painstaking copy editor, and the chapters in this book were improved by her efforts. In addition, she prepared the index. Finally, I am grateful to Dr Sam Elworthy and staff of Auckland University Press for their patience, encouragement and assistance. Sam's and an anonymous reader's comments on the manuscript, and their suggested changes, helped me clarify and strengthen the argument, Anna Hodge, Senior Editor at AUP, provided much appreciated assistance as the manuscript was made ready for publication, and Katrina Duncan designed the book. Any remaining shortcomings are my responsibility.

My thanks go to all these people and to my late grandfather, George Holland, for sparking my interest in the land and landscapes of southern New Zealand.

Peter Holland
Dunedin

The New Land Imagined from Afar, Experienced at First Hand

In 1949, the North American geographer Andrew Clark in his book *The Invasion of New Zealand by People, Plants and Animals* dissected the environmental transformation of the South Island of New Zealand over the previous century by three generations of European settlers:

> There is in the South Island today a lack of any solid rural tradition, of any peas-
> ant-like feeling of love for the land and the countryside . . . . it is this writer's belief
> that the migration of a yeoman-farmer class to the South Island would have
> resulted in an entirely different and much more conservational attitude towards
> the land by its people . . . . People who look on land as a commodity, as a means of
> earning a living which, while different from a factory or a commercial occupa-
> tion, is yet of the same kind as these, seem not to develop a strong resistance to
> practices leading to mutilation of the area in which they live.[1]

I read those stern words while an undergraduate student in Christchurch in the 1960s, and they awakened memories of childhood visits to my paternal grandparents' farm four kilometres north of Waimate in south Canterbury, with its productive flower and vegetable gardens and old fruit trees; small fields of grass, grain and root crops bounded by gorse and broom hedges; clumps of trees; and an air of calm self-sufficiency.

That contrast between memory and scholarly argument established the fundamental questions that have driven my research for two decades and also drive this book. Were the first two generations of European settlers in the South Island of New Zealand uninformed, disinterested people driven by a single-minded urge to transform their recently acquired properties into economically productive farms and sheep stations, regardless of the environmental consequences? Alternatively, did they learn from their experiences in the new land, were they aware of what they were doing to the environment, and did they find guidance for their transformative actions in models developed locally and elsewhere?

There can be little doubt that the first two generations of European settlers fundamentally transformed this country's lowland environments in a very short space of time – the geographer Kenneth Cumberland suggested that what had taken three centuries to achieve in North America had been accomplished within a century in New Zealand[2] – but as I read more about environmental changes in the lowlands and low hill country of southern New Zealand I began to think that Clark had been partial in his analysis. In particular, he wrote little on how settlers learned about the diverse natural environments of their new homes. I suspected that Clark's assessment was rooted in secondary sources of information and aggregated statistics, and wondered if he might have felt differently had he been more aware of early settlers' accounts of the environmental challenges they had to overcome if they were to support themselves and their families in the new land.

This book attempts redress by letting the first two generations of European settlers in the lowlands and low hill country of rural southern New Zealand tell their stories in substantially their own words. I focus on the first four or five decades of organised settlement in the area that lies below 500 metres and extends from north Canterbury to Southland and from the Pacific coast inland to the Southern Alps in order to tackle some key questions about settlers and their environment. What sorts of plants did settlers bring onto their properties to establish field boundaries; feed family members and livestock; provide decoration, fuel and lumber; and give shelter from strong winds and extreme

Motutapu Homestead, Mount Aspiring, 1916. SLATER PHOTO, BOX 194:
HOCKEN COLLECTIONS, UARE TAOKA O HĀKENA, UNIVERSITY OF OTAGO

temperatures to crops and pastures, animals and people? How did they make sense of the South Island's varying weather patterns and the diversity of local soils? How did they recognise and deal with the impact of floods, severe frost and snow, rabbits, thistles and soil degradation? And what were the sources of their environmental knowledge? This book looks at what rural settlers learned about the environments of their farms and stations, where and from whom they obtained environmental information, how they perceived patterns in what they experienced, and what use they made of that information as they transformed their properties.

For the period covered by this book – principally 1840 to 1890, but with extensions into the twentieth century where appropriate – quantitative information about the physical environments of southern New Zealand can be found in official reports to government agencies here and in Great Britain; in published compilations of observations made at meteorological stations in dispersed, mostly coastal, settlements; and in newspaper correspondents' accounts of major weather events and floods. There are also, however, informal accounts of local environmental conditions in the letter books, journals and diaries kept by rural people, and these encouraged me to worm my way into the minds of these long-dead agents of landscape transformation and environmental change in the hope of discerning their goals and expectations, to uncover what they understood about the environmental forces of the new land, and to ascertain whether they had put that knowledge to good use when making decisions about the plant varieties to grow and the animal breeds to raise on their properties.

This book does not paper over the long-term environmental damage that settlers caused in their efforts to develop economically viable operations within a few years. I shall show that most settlers failed to understand that New Zealand is a mosaic of ecologically diverse areas, each primed to deliver surprise after surprise, and that this led to major environmental problems. But I shall also show how the settlers grew adept at detecting and interpreting weather signals, and made economically informed, often scientifically justified, choices when establishing productive pastures for their livestock.

We know something about the larger world of ideas that enabled the settlement and environmental transformation of New Zealand. From the beginning of organised European settlement, the drive to personal and national prosperity was believed to require the application of scientific thinking and advanced technology. In the British Isles, science thrived after the foundation of the Royal Society in the late seventeenth century, and the nineteenth century was notable for major advances in agronomy and biology, communications and transport, engineering and technology. Michael Faraday's public lectures at the Royal Institution in London were popular, and almost everywhere they looked, citizens could see evidence of how science was improving their lives. In New Zealand during the second half of the nineteenth century, that spirit was fostered by several committed individuals: skilled communicators, practising scientists and ardent believers in the power of rational inquiry to foster environmental understanding and lasting economic and social development. They included the biologist and missionary William Colenso, the explorer and geologist Arthur Dobson, the geologist and museum administrator Julius

Haast, the biological and earth scientist James Hector, the geologist Alexander McKay, the geologist and surveyor James McKerrow, and the surveyor and meteorologist Charles Torlesse.[3] To that list we must add the tragic figure of Robert FitzRoy, foundation head of the British Meteorological Office, a former governor of New Zealand and influential in the establishment of a network of meteorological observatories across the country.[4] Shared national and international agronomical good practice, supported by environmental observations and meteorological information, facilitated transformation of the southern New Zealand plains, downlands and low hill country into economically productive land for livestock and crops.

What is less well documented relates to how individual European settlers recognised, learned about and reacted to particular environmental conditions in different parts of New Zealand. Each settler had a unique information field: an expanding mixture of written and spoken sources of environmental and practical observations set in the lived-in landscape. We cannot know what was said in conversations, observed during work and travel, or read in books and newspapers, but we may find sufficient detail in the diaries and correspondence of early settlers to identify their likely sources of environmental information and to model the progression of their learning.

Between late 1841 and mid-1847, for example, the young English settler Joseph Greenwood[5] kept detailed accounts of his daily activities (Figure 0.1). He typically reported the time he spent each day at his places of work: at his brother's small farm in Lowry Bay; in a Wellington shop, for which he also collected money owed by customers; travelling to and from Wellington, Manawatu, Taranaki, Cloudy Bay and coastal Wairarapa to investigate business opportunities; and working on family properties in Port Levy and Motunau in north Canterbury. In his diaries he described his social and community activities: jury duty in the Wellington Court House; attending church services; meeting acquaintances in the street; and visiting neighbours' houses in Wellington, on Banks Peninsula and in north Canterbury. Most of his contacts with Māori came from employing them as guides during his travels in Manawatu and Taranaki, as well as on Banks Peninsula and in north Canterbury, where they thatched farm buildings and assisted with heavy farm work. In some of his Sunday entries, Greenwood referred to walking and reading, but rarely in detail. He did not take a vacation during the six years covered by the diaries, but usually held Sundays, Christmas Day and Good Friday free from work.

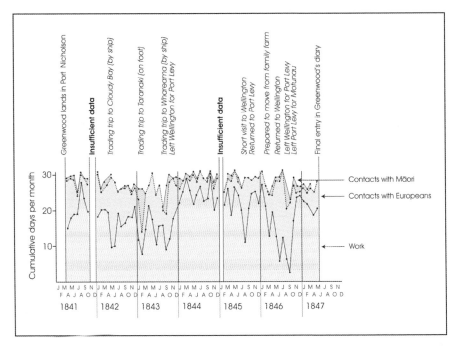

FIGURE 0.1 Joseph Greenwood's theatres of activity and his contacts with Māori and Pākehā in Wellington and Canterbury during the 1840s. SOURCE: INFORMATION EXTRACTED FROM A TRANSCRIPT OF GREENWOOD'S DIARIES; SEE NOTE 5

For Joseph Greenwood, and other settlers, we can examine the changing information field that shaped their understanding of the New Zealand environment, and from Greenwood's diary entries we can begin to map his intellectual landscape. The figure suggests the relative importance of things observed by Greenwood in the course of his work and travels in the two main islands, conversations with and observation of his European neighbours, and contacts with Māori, each expressed as a part of the waking hours in a month. Not every day was described: there were missing diary entries or, more usually, insufficient detail about where he was at the time and with whom he was in contact.

Where did settlers like Greenwood get their information? Local and international newspapers were critical. Widely distributed publications such as the weekly *Canterbury Times* and the *Otago Witness* played an important role in fostering the spirit, if not always the practice, of rational inquiry amongst those rural settlers with the inclination, time, resources and opportunity to rise to

Station homestead, Mount Aspiring. BOX 211: HOCKEN COLLECTIONS, UARE TAOKA O HĀKENA, UNIVERSITY OF OTAGO

the challenge. In the 6 September 1866 issue of the Milton newspaper the *Bruce Herald*, for example, the editor invited farmers in the Tokomairiro district of south Otago to send him information for publication in that newspaper's agricultural column so as to make it 'peculiarly the paper for the country'. Topics relating to the land were of comparable interest to newspaper readers in the second half of the nineteenth century as business news is today.

Settlers also read newspapers from overseas. Even during the two decades prior to the 1860s, well before telegraphic connections and the railway could facilitate dissemination of environmental and other information between settled parts of the colony, rural people were aware of and in frequent contact with a larger world. One early resident of Christchurch, Joseph Munnings, wrote in his diary on 26 May 1862 that he had received copies of the *Calcutta News* and

7

the *Bengal Advertiser* from his sister in India, and on 30 March 1865 recorded: 'To the Mechanics [Institute] in the evening to look over the home [English] papers.'[6] Some commentators were impatient for more information, with the editor of the *Otago Daily Times* on 5 March 1862 decrying the isolation of colonists from the tide of fresh discoveries accessible to people living in England: 'He [the colonist] is forever drawing on the past, and receiving but little from the present.' Nevertheless, settlers exchanged letters and newspapers – including monthly editions of newspapers especially published for posting to the British Isles – with family members and friends in distant places; and provincial centres had an Athenaeum, a mechanics institute or a lending library where members could read imported books, magazines and newspapers. As Annie Wilson, wife of Sir James Wilson of Bulls, wrote to her fearsome father-in-law in Scotland in 1887, 'All colonists are great readers of newspapers, and one's papers pass from hand to hand after we have read them.'[7]

Many towns had at least one commercial printer, and by 1895 some 45 newspapers and fifteen church, union and commercial broadsheets were published in 32 centres between north Canterbury and Invercargill. Editions of most newspapers carried first-person accounts of recent weather conditions and adverse environmental events in different parts of the colony, and a regular reader could find brief items about the American Civil War, battles in Cuba and the Philippines, the progress of Kitchener's troops in the Sudan, or conflict between the British and the Boers in South Africa. In addition, they often published summaries of the findings of recent American and European research into the beneficial effects of guano and mineral fertilisers on crop yields, and information about new pasture plants found by scientific experimentation to be well suited to specific environmental conditions.

The Athenaeums and mechanics institutes where settlers caught up on international news also fostered a lively culture of lectures and debates. Joseph Munnings, the reader of the *Calcutta News* and the *Bengal Advertiser*, attended the Mechanics Institute again on 19 April 1865 and wrote in his diary:

> Went to the Mechanics [Institute] in the evening to hear Mr Doyne lecture on that most interesting subject (to all who reside or take any interest in the province of Canter[bury]) on bridging the Rakaia, showing the superiority of an iron over a wooden bridge, and the utter impossibility of building a wooden bridge over the Rakaia. [He showed] plans of his intended bridge and told us the manner

in which the cast iron cylinders are sunk, and showed us also a new invention of his for the purpose of preventing the [shaking?] of a loaded bridge injuring the iron cylinders. A very interesting lecture, much useful knowledge disclosed, and well [received?]. Mr Doyne received a vote of thanks from all present at the conclusion, when most of those present availed themselves of the opportunity of [looking?] over Mr D's plans of the bridge.[8]

At such lectures, settlers learned much about science and agriculture. James Hector, for example, lectured on 'The Utility of Science' to Dunedin members of the Young Men's Christian Association in 1862:

But exact observation on this matter [the effect of environmental conditions on wool quality] is not merely idle curiosity, but may lead to very practical results by showing the correct way by which the natural pasturage should be nurtured and husbanded, and so enabled to carry a much larger proportion of stock than would at present be possible. The kind of observation is simple, the qualified observers many, and the results very important. Let them set to work, therefore, and gather facts respecting this from year to year and experimenting, if they can, and in time they will reap a harvest of profit from the true scientific methods.[9]

His words must have struck his audience, and the people who read the newspaper report of the lecture, as a clarion call for the application of science to agriculture.

Settlers also learned from books. A clipping from the 20 April 1872 issue of the *Mount Ida Chronicle*, placed by James Preston in his private papers, is a list of books received by the Education Office in Dunedin 'for distribution among the public libraries of the [Otago] province'. It included novels by well-known Victorian writers, accounts of exploration in Africa and the Americas, biographies of important figures in military history and engineering, and regional histories.[10] *On the Origin of Species* by Charles Darwin, three books of popular science and Alexander von Humboldt's *Cosmos*, as well as his account of South America, were amongst them. The published notice concluded, 'In addition to this consignment there is another shortly to arrive.' A few years later, on 31 December 1888, Preston recorded in his diary that on 11 August he had bought 95 books for the newly established Kyeburn Library from Bell and Bradford, booksellers in Dunedin. Most of his purchases were novels, religious tracts, biographies and explorers' tales, but the list included *Cottage Gardening, Live Stock of the Farm, Pleasures of Our Little Poultry Farm* and *The Descent of Man* by Charles Darwin.

At mechanics institutes and Athenaeums, by reading books and newspapers, through travel and in casual conversations, settlers gained knowledge of their new environment. What were they interested in? Weather was a persistent concern for settlers. One of the earliest informal accounts of weather and climate in the South Island was that of John Barnicoat, a surveyor employed by the New Zealand Company. Although based in Nelson, he travelled widely throughout the eastern South Island, and in his diaries[11] often compared the day's weather with what he had known in England or experienced in New Zealand: 'Day [29 June 1842] as delightful as June at home almost.' A contemporary, Thomas Ferens of Waikouaiti in coastal east Otago, wrote in his diary on 20 June 1849, 'Strong SW wind – most terrible gale of a hurricane nature blew all night – with snow, hail and sleet – bitter cold. Quite a transposition of temperature and climate, more resembling the Northern blast of England and Scotland.'[12]

Comparison of local weather conditions with those in the settler's home country was a recurring theme in farm and station diaries at the time. In their diaries, which they kept to send to family in Europe, the Pillans family of Inch Clutha in south Otago[13] recorded this sequence of observations: 'One remarkable feature of the New Zealand weather is that we have seldom or ever the fine warm genial showers so often experienced at home' (26 October 1850); 'On the whole the past month has been fine but the nights have been a great deal colder than what they are at home in the corresponding month which we may call your June' (1 January 1851); and 'We have not had above 2 or 3 days in all of what could be called in Scotland really fine hot summer days' (13 January 1851). On 19 March 1851, Edward Ward[14] wrote in a similar vein to his family in England about the weather in Christchurch: 'Last night had a fresh frosty cold about it – this morning had the same feeling, something of a bracing English autumn, and very pleasant.' One week later, the weather was even better: 'Morning pleasant and cool, very calm; but the heat of the day furious. Many consider this the hottest day we have had since landing. I almost think so – and yet we ought to expect English October weather.' Those conditions persisted into May, when the writer experienced 'The same lovely, most lovely, weather; surely finer weather never was seen at any time of the year in any country . . . .

A lovely day – even at nine o'clock in the morning it was as warm as a summer's day in England and continued fine all day.'

Settlers learned to recognise, although not always accurately, underlying geographical and seasonal variations in the New Zealand weather. On 6 November 1862, for example, the *Otago Witness* reprinted without comment a piece from the 4 September 1862 edition of the *Caledonian Mercury*: 'The climate of the province [of Otago], excepting near the coast where it is fickle and boisterous, is similar to that of the southern parts of England – a little warmer in winter and cooler in summer . . . . frost and snow, except on the higher ranges, are comparatively unknown . . . . [and] generally speaking, the soil is the same on the tops of the ranges as in the adjacent lowlands.' Three weeks later, on 23 September 1862, the *Otago Daily Times* glossed a manual that had been produced for the information of newly arrived goldminers in which the climate of Otago was described as mild, dry in winter and with snow on the mountains. 'In winter, also, the waters of the Clutha [River] are at their lowest', rendering the riverbed accessible for sluicing. Published comparisons with European weather continued throughout the nineteenth century, albeit less frequently, at a slower pace and a little more accurately, with a member of the McMaster family of north Otago[15] recording 'The warmest day [6 January 1886] I have yet experienced in New Zealand. Thermometer 103 [° Fahrenheit] in sun and 90 in shade.' Such statements steadily gave way to local comparisons, like this one from the Ida Valley Station diaries[16] on 9 January 1899: 'Terribly hard frost last night and all today – by far the hardest this winter, as hard as I can remember.'

During the second half of the nineteenth century, European settlers' search for rational understanding of the environment must have been sorely tested by episodes of flood and drought, galeforce winds, frozen topsoils and severe snowstorms. But landholders were encouraged by the newspapers and magazines they read to seek environmentally appropriate and economically effective ways to develop their rural properties. On 20 February 1868, for example, the *Bruce Herald* published an original article warning local farmers about the environmentally deleterious effects of burning-off in the extensive tussock grasslands of south Otago. Six months later, on 29 August, the editor followed up with a trenchant criticism of farmers near Milton who had set fire to wheat stubble after the grain harvest: 'It would appear that the settlers who are guilty of such an extraordinarily foolish act do not know that the soil of this plain [Tokomairiro] has no super abundance of vegetable matter, and consequently its powers of fertility are the reverse of inexhaustible.' He was not the only public figure then or later in the nineteenth century to voice concern about the limitations of the New Zealand environment.

The rapidly declining area under native forest, mounting cost of imported wood for construction and fencing, steady reductions in soil fertility and crop yields, and worsening problems with weedy plants and pest animals were frequently debated in the media and parliament by the end of the nineteenth century. Settlers came to realise that the environments of southern New Zealand were not as uniformly benign as they had been led to believe, although many landholders were tested even more by macro-economic difficulties. While not especially evident early in the period of organised settlement, environmental problems were more frequently reported later in the century, and for many residents they diminished the country's attractive features. This was noted by several commentators, one of whom, the anonymous 'Hopeful', described the New Zealand climate as 'boisterous, ramping and uncertain', and 'a poor one after England'.[17]

Amongst the unexpected difficulties were severe storms of rain or snow that, depending on the season, affected the area during most decades and were frequently associated with widespread flooding. But the sorest disappointments arose from plants and animals accidentally or purposefully imported by well-meaning plant merchants, individuals and acclimatisation societies to improve settlers' quality of life: small birds such as sparrows and linnets to control insect pests and bring a touch of home to the farm or station, gorse and broom for hedges, trout and salmon to stock rivers, rabbits for recreation and food. The list is very long, and a substantial proportion of the newcomers thrived in southern New Zealand, spread from their places of introduction and caused environmental problems. The issues are complex, but they stemmed from people importing known species to create new productive ecosystems, and then discovering that the environment could not contain, let alone constrain, several of these newcomers. This story could be told by reference to several adverse environmental conditions as well as numerous weedy plant and pest animal species, but the availability and coverage of contemporary eye-witness accounts limited my choice. For that reason, I shall refer to too much or too little precipitation, prolonged episodes of severe frost and deep accumulations of snow, the spread of rabbits across the landscape, the occupation by gorse, thistles and other common weeds of cultivated ground, and declining soil fertility.

Until the 1880s, environmental inquiry was largely the purview of interested amateurs. Nevertheless, rural people learned to recognise then interpret key environmental signals, and they employed those skills in their daily lives. One contemporary aspiration was for rural people to view each farm as a laboratory, each field as an experiment and each farmer as a scientist. Establishment of the New Zealand Department of Agriculture towards the end of the nineteenth century facilitated the professionalisation of agricultural and pastoral science in this country, and farmers were encouraged to seek practical advice and useful information from the cadre of scientists and advisors employed by the state. This shift in emphasis, as discussed in later chapters, was not without adverse consequences for this country's lowland rural environments.

A SOUTHERN LABORATORY

The landscapes of southern New Zealand to which the first generation of European settlers came were commonly viewed as a melange of high mountains, steep slopes and broad plains; prone to flooding; covered by large and small patches of forest set in broad expanses of tussock grass, shrubland and marshy ground; possessing readily cultivated and apparently fertile soils; with a fairly low density of Māori; and home to insects, birds and plants quite unlike those that newcomers might have known in Britain. Little wonder, then, that they believed it was their personal responsibility to transform the antipodean wilderness into God's garden. Did they recognise the paradox? If God had created the living things that they had known and nurtured in Europe, did He not also create native New Zealand plants and animals and were they not deserving of their care? Even while settlers were clearing large tracts and small relic stands of native forest with axe, saw and fire, they were planting vast numbers of imported tree and shrub species for shelter, fuel, lumber and decoration. Rural people shot native forest and wetland birds for food and sport, yet kept kākā as household pets. In the 1820s, Charles Darwin had come to a bleak assessment about the native plants of New Zealand, believing them likely to succumb once the trickle of imported plants had become a wave and many had naturalised. Thinking of that kind was more received truth than observationally justified opinion, and it is evident in the following words of Edward Ward, written on Sunday 12 January 1851, after viewing the high plains of mid-Canterbury from the foothills of the Southern Alps near Oxford:

A mile of walking from the Sitz [the name he gave his camp site] took us to
Captain Mitchell's station, planted in rather a picturesque, though rather dreary
position, just under the foot of Mount Grey, in a valley by the side of which is a full
river in winter time. It consists of one slab house merely, but it was refreshing
to see any sign of life at all after the weary lifelessness of the great [Canterbury]
plain. Even the cows at a distance, wandering about, gave it an English charm,
and the whole went, along with the other mere dots of cultivation over the country which we have seen [so far], to show what may be done and how magnificent
the whole will appear when the tide of life runs full over what has lain unoccupied so long.[18]

Ward saw what might be, but the clergyman James Stack of Kaiapoi, who
accompanied Bishop Harper on his tours of the eastern South Island from
Christchurch to Invercargill between 1858 and 1860, described the landscapes
of southern New Zealand as they were: 'Looking southwards [from high on the
Port Hills] an apparently boundless plain stretched away from our feet as level
as the sea, of one uniform colour and one uniform covering of yellow grass.'[19] He
also reported the trying conditions endured by the earliest European settlers
on the Canterbury Plains, who were then living in small, poorly constructed
huts set in a sea of tussock far from the nearest neighbour, and wrote this telling
account about the onset of a southerly buster:

A small but very dark cloud appeared above the southern horizon . . . in less than
thirty minutes it covered half the sky . . . . Suddenly, with the violence of a whirl-
wind, the southwest gale burst upon us, enveloping us in a blinding cloud of dust,
dry grass and twigs . . . . Instantly, the temperature fell many degrees . . . . the rain
came down in torrents, followed by a violent hailstorm . . . . Every landmark that
could guide us was hidden from view . . . . There was no likelihood of the storm
abating for twenty-four hours.

The agricultural and pastoral plains, downlands and low hill country of
southern New Zealand are good places in which to explore the nature and progress of environmental learning, as well as the intensity and timing of landscape
transformation, in settler society. The extensive tracts of tussock grassland
and low shrubby vegetation first seen by European settlers in the 1840s and
1850s were readily cleared and their soils cultivated. Even the many large and

small areas of wetland did little to impede the spread of European settlement. In the late 1840s, when organised settlement began, the area was occupied by few Māori and, unlike the situation in other parts of the colony, most environmental forces operated within clearly delineated topographical boundaries. In their understandable desire to ensure a steady flow of new residents, the New Zealand Company and other such organisations published handbooks for intending settlers. As well as manifestly sound advice, however, they contained statements that wittingly or unintentionally misrepresented environmental conditions in the young colony.[20] Despite its geographical latitude, New Zealand does not experience a Mediterranean climate. Rather, its weather systems and climate reflect the country's geographical position at the centre of the ocean hemisphere as well as the prevalence of mid-latitudinal westerlies. The territory's north-to-south spread, spanning almost fifteen degrees of latitude, ensures that several distinctive weather systems will be regularly experienced, and the mountain backbone enhances their variability. An early lesson learned by European settlers was that, despite what they might have read before embarking for the new land, there is more than one climate in New Zealand. More importantly, while environmental diversity may be the overriding current perception of this country, this was not manifest in published handbooks for settlers until the New Zealand government produced its own material for the information of immigrants.

European settlers had it mostly their own way until the late 1860s. There had been periods of less than optimal rainfall, widespread flooding and outbreaks of polar air that brought episodes of savage weather to the eastern and southern South Island, but farmers seldom reported severe erosion on their properties and were still achieving good wool clips and satisfactory crop yields. By the middle of the 1870s, however, rabbits and flocks of seed-eating birds had become significant pests, gorse was showing its invasive tendencies, and herbaceous weeds such as Californian and Scotch thistle as well as fat hen and sheep sorrel were reducing economic yields on farmland.[21] To add to their mounting environmental concerns, farmers had to respond to declining soil fertility by top-dressing with mulches and fertilisers during a period of poor export returns for primary products. The image of England's farm in the antipodes comforted and inspired five generations of rural New Zealanders, but as a model for environmental transformation it was to have undesirable implications for this small, geographically isolated country.

CHAPTER ONE

Māori Environmental Knowledge
An Imperfectly Realised Resource

The early waves of European settlers in southern New Zealand did not arrive in an unoccupied land. For most of the newcomers, ignorance of Māori language and customs was a severe constraint on the flow of environmental information, but differences between the two world views, combined with the European settlers' reliance on their own knowledge of scientific agriculture and pastoral farming, were even greater impediments. As a result, what could have been a useful resource for improvement was not extensively used by Pākehā.

The first Māori settlers were undoubtedly challenged by the diverse physical environments of Te Wai Pounamu, and the lessons they learned left them well placed to inform the first generation of European settlers about the topographic diversity, environmental conditions and hazards of southern New Zealand.[1] Amongst Māori, the holding and transmission of valued knowledge was largely formalised: it was retained by carefully selected, highly trained individuals on behalf of whānau and hapū, and orally transmitted to those with the right to have it. James Stack understood this well. 'All boys of Rangatira rank were obliged to attend the classes taught during the winter months in the Wharekura by individuals learned in History, Mythology and the various branches of knowledge possessed by the Maoris . . . . The lessons were difficult, and the discipline severe.'[2] Tohunga 'would often pass long hours of the night in contemplation of the stars, and would be looked upon as reliable weather prophets. Travellers and fishermen would consult them ere venturing forth, and their powers are said to have also enabled them to foretell the general aspect of coming seasons, their fruitfulness or otherwise.'[3] Elsdon Best reported the significance of weather forecasting by Māori through their close observation of heat shimmer and clouds, as well as the form of a crescent moon, the appearance of Pareārau (possibly Saturn or Jupiter), the Magellanic Clouds, Rehua (Antares in the constellation Scorpio), Orion, Rigel as it rose above the horizon, the Milky Way and Matariki (the Pleiades). He also found that Māori thought abstractly about the physical world.[4]

Although communication of information about tapu native plants and animals as well as mahika kai (food gathering places) was closely controlled in traditional Māori society, there were few restrictions on telling strangers about trails, landmarks and topography.[5] Nevertheless, settlers during the first two decades of organised settlement in southern New Zealand effectively chose to learn for themselves, repeating earlier errors and making changes to the environments of the south that would have been better avoided. For most settlers their only exposure to the environmental knowledge of tangata whenua came from reading occasional newspaper pieces. In marked contrast to Māori, the transmission and storage of environmental knowledge in early settler society were haphazard, and tended to be restricted by cost and other economic factors, as well as technical, geographic or linguistic impediments to communication.

Relatively few Europeans became fluent in te reo Māori or learned directly from iwi. While this was partly due to the small numbers of Māori living in coastal settlements, at the same time Māori shearers were ranging across the South Island and making contact with European settlers along the way. During the 1860s the Māori settlement at Kaikoura 'supplied labour to the stations at short notice, as many as sixteen Maoris having been known to bivouac at Hawkswood [Station] at the same time';[6] and George Levens, who farmed at Willowbank near Geraldine in south Canterbury, reportedly said that 'When we arrived here, there were about 500 Maoris in the [Arowhenua] pa, and we had to learn their language in order to get them to work for us.'[7] A further reason for the disjunction between Māori as repository of local knowledge and settlers related to the linguistic skills of European administrators, functionaries and scholars. While some, such as Elsdon Best, William Colenso, Walter Mantell and Edward Shortland, showed an awareness of Māori environmental knowledge and protocol, and were reasonably fluent in te reo Māori, they were few and far between. However, the principal reason for the relatively small amount of environmental information that passed from Māori to Pākehā related to the lower standing in settler society of indigenous environmental knowledge compared with that stemming from western science. That there was at least some communication between Māori and Pākehā is evident in the obituary for Horomona Taupiki, known as 'Solomon the Maori', published in *The Press* on 3 November 1863: 'A remarkably shrewd and intelligent young man, and speaking English better than most of his countrymen, he was able to give much useful information to the pakeha.' He drowned while escorting a European settler through the back country. Despite the impediments, references to Māori environmental knowledge were published in newspapers and magazines, and recorded in diaries, letters and other private papers.

In the following narrative, several individuals stand out: eminent people like Huruhuru, who told Shortland about diurnal and seasonal variations in the discharge of the Waitaki River and gave his European guest a hand-drawn map of southern New Zealand; Edward Shortland, who published the findings of his exploration along the coastal strip between Christchurch and Invercargill; and Maori Jack, who showed settlers where they could cross the Alps between Otago and the West Coast. Early in the period of organised settlement, Māori were asked about passes between Nelson and Canterbury via Marlborough, between mid-Canterbury and the West Coast goldfields, and from Otago to the West Coast. The first overland routes allowed settlers to drive large flocks of sheep from Nelson and Marlborough into the tussock grasslands and low shrub country of Canterbury and north Otago, and thence to the inland basins of the South Island. Some European settlers also believed that Māori knew about historical episodes of unusual weather and stream discharge – particularly heavy rain, gale, flood, snow and drought – and recognised signs of adverse weather.

EXPLORING PLACES AND SPACES

Herries Beattie, an avid collector of Māori knowledge and lore, documented the navigational skills of tohuka;[8] Bawden, a social historian, reported that

Māori had told Marsden there was not safe passage for an ocean-going ship between the Cavalli Islands and the mainland;[9] and one J. L. Nicholas found that 'If the star they [Māori navigators] look for does not appear at the time it is expected to be seen, they become extremely solicitous about the cause of its absence, and immediately relate the traditions which they have received from the priests concerning it.'[10] At about the same time, Lieutenant Cruise observed that when Māori voyaging in the *Dromedary* from Sydney to Northland were 'asked at any hour of the day, where their country was situated, they pointed to the east with the accuracy of a compass; and when the stars appeared in the evening, they displayed equal sagacity'.[11] When the *Dromedary* had passed the Cavalli Islands and reached the Bay of Islands, 'The delight of the New Zealanders, as they saw successively the different parts of the country with which they were familiar, was excessive; they ran up the rigging with the activity of seamen, shouting the names of the various head-lands.'[12] In a statement to the House of Lords in 1838, J. S. Pollack referred to the navigational and other maritime skills of Māori: 'If they get a good captain they do very well. There are many employed by the Americans as well as by us; many are boat-steerers in the American vessels.'[13]

Māori locational and directional skills on land were of a comparably high order. In 1849, in a report on the Nelson Settlement to the New Zealand Company, Bell wrote: 'The Maoris had told [an unnamed official in the New Zealand Company] of a track through the Wairau Gorge by which the Rangitane made their escape from Te Rauparaha.'[14] Farther south, it was inferred that Māori had 'made frequent excursions into what is now known as the Mackenzie Country', for Huruhuru informed Shortland about the best routes into the interior, good stopping places along the way, and places in the extensive inland basin where he would find eels and weka in abundance for food.[15] Overland travel in much of the South Island meant having appropriate footwear and being able to cross large and small rivers safely. To those ends, Shortland and his party were ferried across the Waitaki River on raupō and flax rafts,[16] given torua (sandals woven by local Māori from the stiff leaves of tī or harakeke that Shortland found better than leather-soled boots on the rough gravels of the south Canterbury coast),[17] and shown how groups of five or six men should stand side by side with their upper arms looped over a long stick whilst fording a river in spate.

Through his skill as a linguist, Shortland learned that the Māori chief Totaranui had informed Captain Cook that the new land comprised three main islands. He also reported that 'Huruhuru's leisure in the evenings was employed in giving me information about the interior of this part of the [South Is]land, with which he was well acquainted.'[18] The considerable geographical and environmental knowledge of Māori, and their interest in acquiring more, impressed many early nineteenth-century visitors, one of whom, Lieutenant Philip King, noted that Chief Te Pahi of Northland 'never missed a chance to gain new knowledge'.[19] During his travels in southern New Zealand, Shortland overheard conversations between old Māori men about places where eel and weka could be caught in the open expanses of the Canterbury Plains, was told by his Māori guides about the grass and shrub lands of the Mackenzie Country far to the west, and was impressed that they had names for the large and small topographic features of areas close to the coast through which his party had walked.[20] He believed that 'Those who are anxious to teach the New Zealanders English will be better able to do so having first learned their language.'[21]

Other company and government officers had comparable experiences. During his stay in Otago, the New Zealand Company surveyor, Charles Kettle, was told by Māori about the interior of Te Wai Pounamu,[22] and Huruhuru rafted Bishop Selwyn across the Waitaki River, which the latter incorrectly described as 'milk white with melted snow'.[23] Towards the close of the 1840s, Thomas Brunner travelled extensively on the West Coast 'accompanied only by a few unenthusiastic Maori and which he survived by the eating of fern root and the wearing of sandals of Phormium [flax]'.[24] Recent arrivals, like the young Englishman, Thomas Ferens of Waikouaiti, made good use of the geographical knowledge of local Māori, and on 2 February 1849 he recorded with evident relief how he had been accompanied by 'a Maori on horseback to town [Dunedin]. When on the Snowy Mountains [the flanks of Mount Cargill] on the NE side a dense fog came streaming up upon us.'[25]

Amongst the earliest records of Māori assistance to Europeans interested in pioneering trails across the mountains that would be suitable for people and livestock were the help provided by Reko to John Thomson in traversing the Nokomai–Nevis route[26] and the advice given to Frederick Weld. In his report published by the *Lyttelton Times* on 8 March 1851, Weld described how he had followed

the ridge of a hill along the worn channel of a deserted native path, famous in the wars of old, [and] descended into the Tutaiputuputu. We had bidden farewell to the prairies of the South, and were evidently journeying amongst the offspurs of the Kaikouras. [At the whaling port of Amuri we found] two large boats hauled up belonging to natives who were on their way, by easy stages, from Motueka to Lyttelton to work on the roads; one old man, the chief, said he was going there to die in the country of his fathers. [At the pā of the Ngāi Tahu chief, Kaikoura, our party was given] a hearty breakfast. I questioned him on the subject of inland communications with the Awatere and Wairau, and found that I had been right the previous day in my surmises. He described a pass in the mountains to be reached by ascending the Tuahuka, and said that formerly he had often been there to catch kakapos in a black birch wood above the Awatere pass. As I could here have obtained provisions, I was at first tempted to retrace my steps and attempt to find the pass, but he alone of all the natives knew it, and none of them could I engage to carry provisions or accompany me. [There is a ridge that] separates the Awatere from the Waihopai, over which there is here an old native route . . . . The pass, though high, did not look impracticable for stock, and I regretted that time did not allow me to ascend it. [He subsequently observed an opening in the hills flanking the Waiau-ua leading into the Amuri.] This probably is one of the two old native passes: the double pass and the black birch kakapo bush between the Clarence and the Awatere mentioned to me by Kaikoura, must be to the east of Barefell's Pass where the off-spurs of the Kaikouras cover more ground and are higher, yet I doubt not could be penetrated by following some branch of the Awatere.

Similar use of Māori geographical knowledge was evident in other records. In the south, during the 1840s and early 1850s, first local whalers then European settlers used Māori tracks running inland from Caroline Bay, several of which were later surveyed for streets in the south Canterbury centre of Timaru. '[Settlers] moved with difficulty through virgin country thickly covered with a mat of vegetation, finding here and there short stretches of track made by the Maoris, which gave them relief from the spines of the wild Irishman and the piercing spines of the Spaniard.'[27] A short time later T. S. Mannering cut a bush track along the line of an old Māori trail to drive sheep between Snowdale and Birch Hill stations.[28] Use of such knowledge was also evident in the layout of large pastoral properties in south Canterbury. 'All runs were to be rectangular in shape and Maori names of objects, points, rivers, etc., intended to define the boundaries, were to be used as others led to confusion and disputes.'[29] In 1858 James Mackay and three Māori guides walked the 200 kilometres from Nelson to Mawheraiti on the West Coast, thereby demonstrating that cattle could be driven from Nelson to the goldfields;[30] and for the North Island Charles Hursthouse cited the work of the geologist C. Forbes, who had been told about traditional Māori knowledge of a sea-level link between Evans Bay and Lyall Bay, Wellington.[31] These services were not necessarily free, as Henry Sewell discovered when the residents of a pā on Banks Peninsula offered to show him the path to Akaroa 'for a consideration'.[32] G. W. H. Lee – who was instrumental in opening up the stock route from Marlborough to Canterbury – and two Māori guides visited the Lake Brunner area where, according to a short article in the 17 April 1858 issue of the *Nelson Examiner*, his travelling companions reported finding quantities of gold in the area. At the time 'the only means of crossing [the Wairau River at Blenheim] was [by] a Maori canoe which is scarcely more than a plank, not being hollowed more than three or four inches in the inside .... The Maoris charge 1/- for a passenger.'[33]

That decade, Arthur Dobson had learned te reo Māori from a 'half-caste' who knew a little English, aided by a copy of the New Testament in Māori and three weeks spent in a pā, in order to facilitate his work as a surveyor.[34] He described his Māori field assistants as 'ideal bush men', 'very good chainmen' and 'quick to learn', and amongst the items of equipment he adopted from them were the kawe (two long bands of plaited flax for backpacking heavy loads), flax sandals for traversing riverbeds and beaches, long poles for crossing rivers, flax-stalk mohiki that served in the absence of canoes, flexible ladders made from flax and thick vines for negotiating cliffs and bluffs, and whata to keep food out of the reach of rats.[35] Dobson published a report of his findings in *The Press* on 2 March 1864 and in it described the difficult terrain he had traversed as well as how Māori used micaceous sandstone to cut greenstone found as boulders in glacial deposits.

Of greater interest to Dobson's readers, however, was his statement that 'Three passes were used by the Maoris in going from the Canterbury Plains to the West Coast, the Taramakau, the Arahura and the Hokitika. No white man has as yet to my knowledge crossed the Arahura Saddle, but I think it must be a bad saddle or the Maoris would have used it more.' During the West Coast gold-rush, the possibility of land access between the east and west coasts of the South Island was high in settlers' minds, and the existence of such passes was of great interest to them. On 8 February 1865 the West Coast correspondent of *The Press* reported that 'One of the storekeepers has ascended the Arahura [River valley] with a Maori to try and find a road to the Six-mile', and a brief account published by the same newspaper on 24 March 1865 described J. S. Browning's surveying trip from the head of the Waimakariri River to the West Coast goldfields via a pass 'used in olden times by the Maoris . . . . [and] the same as that which was pointed out by Mr Torlesse in a letter to the Lyttelton Times the other day'. On 8 April 1865 the editor of *The Press* referred to information obtained from Māori about the headwaters of the Rakaia River, a day and a half by foot from either the Arahura goldfield or Hokitika. Five days later the editor concluded that the most suitable route from Christchurch to the West Coast was that explored by Arthur Dobson, involving the Bealey River and the Otira Gorge, 'unless, indeed, the Maori tradition respecting a pass at the head of the Rakaia turns out to be authentic, in which case if a road can be constructed there it will no doubt form the most direct and advantageous route'. Even so, on 15 April the editor reported that 'Mr Harman has with him the Maori instructions for finding the route [from the head of the Wilberforce River across the Alps] and the map drawn by Maoris'. Harman, who had apparently been told about mountain passes known to Māori by a surveyor and government official, James Stack of Kaiapoi, wrote to the Honourable John Hall about possible land routes between Canterbury and the West Coast goldfields, and his letter was published by *The Press*. He identified Arthur's Pass as a suitable candidate but suspected the presence of other alpine passes in the headwaters of the Rakaia River, which led the editor to conclude on 1 May that 'the Maori pass has not yet been discovered'. Even so, the topic remained of great public interest, and on 5 June 1865 *The Press* reported a rumour that a party of goldminers had found a cave at the head of an alpine pass between Canterbury and the West Coast and wondered if it might be the key to the presumed Māori pass. A week later, Harman wrote to the editor to urge the release of Browning from his mapping duties so that he could survey the route followed by Browning and fellow explorer Griffiths across the Alps, 'which may be the same as the sought-after Māori Pass'. On 21 June 1865 *The Press* concluded that Griffiths' description of 'the Old Maori Pass' linking Hokitika and Christchurch via the headwaters of the Wilberforce, now known as the Browning Pass, conformed with Māori tradition.

Later that year the search shifted southwards, and on 9 August 1865 *The Press* reprinted a piece first published in the 5 August issue of the *Timaru Herald* about a reward offered to the discoverer of 'a pass through the Mackenzie Country to the West Coast . . . . We believe that the Maoris told

Captain Gibson of the existence of such a pass when he was on the West Coast.'
Three days later *The Press* reprinted another short piece from the *Timaru
Herald* about a possible pass between inland south Canterbury and the West
Coast via the headwaters of a river draining into Lake Ohau. Dr Julius Haast
had surveyed the headwaters of the Waitaki River system in 1862 and in his
published report, summarised by *The Press* on 4 September 1865, wrote 'there
is no prospect of finding a practicable pass by the headwaters of the Waitaki' to
the West Coast.

In its 29 May 1865 issue, the *Otago Daily Times* reported that 'Mr O'Neill of
the Survey Department was directed by the Government to endeavour to find
a practicable track from the head of the Lake [Wanaka] to the West Coast [but
was unsuccessful because] he took the wrong side of the river and did not follow,
or lost, a known Maori track.' This news was reprinted in *The Press* three days
later. Elsewhere in Otago, a government-sponsored exploring party set out
from the Dunstan goldfield to search for a pass running from the head of Lake
Wanaka to the West Coast. On 8 September 1865 *The Press* reported that this
group included Mr Vincent Pyke, Mr Coates, three hired men and a Māori guide,
who hoped to discover 'the supposed Maori track from the head of Lake Wanaka
to the West Canterbury Goldfields'. The venture's success was reported in the 18
October issue of the *Dunstan Times* and six days later in *The Press*.

On 2 May 1868 the *Otago Daily Times* drew upon a piece originally published in the *Oamaru Times* about a four-day journey from the Haast diggings to
Lake Wanaka, describing it as a possible cattle track and 'pretty easy all the way'.

We may also mention, as a piece of intelligence, that Maori Jack, who acted as
a guide to Mr Pyke on his journey, has been to the West Coast by what he con-
siders a more practicable route than any yet discovered. He proceeded up the
Matukituki River to Mount Aspiring, which he kept to his right. He found an easy
pass from there. Maori Jack is the first man who has accomplished this journey
in that direction. Dr Hector, we believe, attempted it, but failed. The route is said
to be a shorter one than by way of Makarora.

The flurry of public interest in presumed alpine passes known to Māori
died down almost as quickly as it had begun, but Māori topographic knowledge remained a desirable commodity for settlers throughout New Zealand.
Three significant examples of this are: information about a reef of hard rock

above Brown's Ferry for a bridge across the Waitaki River;[36] the rocks off Cape Saunders at the tip of the Otago Peninsula over which Māori had observed the sea always broke during a southeast gale, as reported in the *Otago Witness* on 15 September 1868; and the point on the Wairarapa coast where Māori had once launched their canoes that was later used by a local landowning family as a safe place from which to launch surfboats to ferry wool packs out to waiting ships and to unload supplies for the station.[37]

WEATHER AND CLIMATE

Weather and climate were topics of vital concern to European settlers, several of whom directly or indirectly sought information about weather systems and flooding from tangata whenua. An early documentary source for Māori weather knowledge is the run of diaries kept by John Barnicoat, a surveyor employed by the New Zealand Company and based in Nelson.[38] 'Another day of heavy rain with occasional intermissions. The natives told us yesterday – "Rain tomorrow, no rain next day".' One day later he wrote: 'A fine day according to the predictions of the natives, who as in other instances have shown themselves remarkably weather wise. A small shower or two in the course of the day.' A few months later he reported: 'Another and third day [12 June] of rain. Nelson is it seems in a great measure under water. It appears that the natives tell us that we are to have no fine weather till Tuesday next, after which we are to have a fortnight of fine days and we are to expect three or four such days as this each fortnight at the change of the moon.' Two days later he wrote, 'A fine cheerful day, thus verifying so far the prediction of the natives. The hills are again covered with snow, having been left there after the late heavy rains.' Then, on 6 October 1842, 'We hope that the rainy season may now be over, but we are told that the Maories say it will continue another moon.' The information he received was not always reliable, as this entry dated 13 March 1844 shows: 'One [Maori] said there would be no wind, another that there would be too much, there would be rain all day, there would be head wind, there would be a heavy sea, etc. We left [Nelson] in a calm, after a few hours sail it blew a gale and veered round right ahead [off Takaka].'

The following oral tradition was collected by Johannes Andersen in 1897 from Māori living on the banks of the Makawhio River in south Westland:

They said that
The old Maori tell by stars and clouds
Easterly wind land breeze good weather
Red sky at night good weather.[39]

Andersen was also told by Māori in the area that seasonal weather could be foretold from observations of the Pleiades and other star clusters, and that if Aoraki (Mount Cook) was kura (a particularly bright red) at dawn then bad weather would soon follow and canoes should not be launched. In Canterbury, a Māori belief recorded by historian James Cowan was that if a northwest arch of clear sky over the Rangitata Gorge was 'ill-formed', then the wind from that quarter would soon give way to stormy weather from the south.[40] Cowan also described how Māori recognised that storms moving in a northeast direction up the Pacific coast could be deflected eastwards by Banks Peninsula, the colour of the cloud bank and its speed of approach indicating the likelihood of that happening. In another tradition involving weather indicators, Cowan recorded that during the hours before a major flood around 1860, Māori were caught by surprise when certain birds (possibly wood pigeons) did not disappear as they usually did before heavy rain. In the back country of south Canterbury the combination of smooth light-grey clouds, air temperatures warmer than usual before rain and weka going to ground during the winter months was believed by Māori to indicate that snow was on the way. One of Cowan's informants told him that Pākehā did not trust Māori readings of weather indicators. On one occasion, when he was employed as a musterer in the Mackenzie Country, his Pākehā companions died in the snow after they refused to follow his lead and leave the area before the onset of what proved to be an unusually severe winter storm.

In traditional Māori society, birds and plants were appreciated as weather indicators: flocks of mohua rising from and then falling back to the forest canopy warned of an approaching storm; how riroriro oriented their nests, as well as how high they built them in the trees, indicated seasonal weather;[41] and on 12 December 1876 a short piece in the *Otago Daily Times* reported the first call of the migratory pīpīwharauroa, which was believed to mark the start of an Otago summer. It was also thought that the position in the canopy of the first kōwhai flowers in spring indicated likely weather conditions during the forthcoming growing season.

In 1823 Lieutenant Cruise of the *Dromedary* reported: 'The day after we arrived, one of the natives whom we had brought round from the Bay of Islands announced his intention of leaving us. This man called himself the priest and pilot of Shukehanga [Hokianga], and was supposed by his tribe to have power over the wind and waves; an influence which, when he was asked to exert it during the late gale, he declined by saying, that he could not do so in the Dromedary, but if he were in his own canoe, at his word, the storm would instantly abate.'[42] Other claims seem more plausible, such as the indications of drought reported by Mr Reay to the Reverend Mr Saxton;[43] or Huruhuru's advice to Shortland to build overnight shelter facing northeast – by nightfall the wind would usually be from the westerly quadrant – so that smoke from the fire would not be blown into the shelter. Settlers correctly believed that Māori recalled major weather events and their consequences for local people. Francis Pillans, a recent settler at Inch Clutha, recorded in his diary on 17 January 1851 that 'From what the Maoris say, it is doubtful if such a flood as this has occurred within their memory.'[44] In the second edition of his book, Charles Hursthouse indicated that Māori also maintained abstract lore: 'The natives, however, consider August the first spring month, and May the first winter month; and this division, as regards spring for the North Island at least, is perhaps the best.'[45] He added that a Māori month does not start at the same time each year, the reason being that the Māori calendar is lunar.

Calamitous weather during the late 1860s renewed European settlers' interest in indigenous weather lore. Lady Mary Barker wrote about the July–August 1867 snowstorm that had been experienced across much of the eastern South Island: 'Maoris are strong in weather traditions, and though they prophesied this one, it is said that they have no legend of anything like it ever having happened.'[46] In a later book she described the years that she and her husband had spent in the foothills of mid-Canterbury: 'We had nothing to go by except the Maori traditions, which held no record of anything the least like that snowstorm. Indeed, I had seldom seen snow lie on the ground for more than an hour after the sun rose, and it never was thought of as a danger in our comparatively low hills.'[47]

The words of seasoned newspaper reporters, like these from the 6 January 1868 issue of the *West Coast Times*, convey much the same story: 'There are Maori traditions which tell us, that at certain intervals, there come and go years without a West Coast summer. We can believe the truth of the legend.' On 9 June the previous year an unnamed resident of Switzers in northern Southland wrote to the editor of the *Lake Wakatip Mail*, published in Queenstown: 'The winter has set in with a vengeance. We have had a snow storm that has continued four days, blocking up the roads, stopping traffic, and far worse than this seriously impeding mining operations. I now begin to believe the Maori prophecy of August 1867 – that New Zealand will be visited by a succession of storms and rain alternately and continuously for one year – will be fulfilled.' An editorial in the 15 February 1868 edition of the *Otago Witness*, presumably inspired by the widespread flooding in eastern and southern New Zealand earlier that month, picked up this theme: 'From remote Maori tradition, the record of periodical heavy storms, about one in ten years, we believe, has been handed down .... There is no doubt this very flood has been predicted.' Farther north, Māori at Arowhenua had apparently predicted a second flood following closely that of 2–3 February 1868, but it did not eventuate.[48]

Throughout the second half of the nineteenth century, Māori weather lore was published in daily newspapers, such as this piece from the 9 January 1886 issue of the *Otago Witness*:

> It might be interesting to learn how the Maoris were able to predict that the season would prove a dry one, as it may be remembered they did, from something they saw in the flower of the flax. That their prediction has been correct is well known to farmers and graziers, and is borne out by the records of rainfall for December where they are kept. In Wellington, for instance, the rainfall for the past month was barely above one inch, as against 12 inches recorded for December of the previous year.

A similar report was printed in the Agricultural and Pastoral News section of the 3 July 1896 issue of that newspaper:

> Maori observers of the seasons prognosticate for us an early spring, and those of the natives who cultivate the ground have already got it in order for spring crops. The natives anticipate an early return of the whitebait to the rivers. Also they put forth the prospect that the spring will be accompanied with an abundance of blossom on the fruit trees. Altogether the Maori weather prophet gives a healthy forecast of the coming season.

COASTS AND RIVERS

By the end of the nineteenth century several Māori traditions of tsunami had been published. Andersen had been told about a tidal wave on the Pacific coast, which one informant thought had affected Blueskin Bay a little to the north of the Otago Peninsula.[49] Another informant identified Moeraki in northeastern Otago.[50] In a further instance, the master of the schooner *Rifleman* reported to the *Otago Daily Times* that Māori living in the Chatham Islands 'were in a great state of alarm' and had retreated inland for fear of a repeat of the tsunami that had struck the east coast of the South Island on 22 August 1868.

Traditional knowledge about variations in river discharge was reported by Shortland: 'Tarawhata assured us that, in the summer season, there is a manifest difference in the depth of water in the [Rangitata River] evening and morning, that of the latter being shallowest; and he supposed this difference to be due to the greater quantity of snow melted during the day, which would have arrived thus far by evening, but would be drained off by the morning.'[51] Tarawhata's explanation doesn't square with modern hydrological knowledge, but during the nineteenth century this belief was sufficiently strong amongst settlers for John Grigg of Longbeach Station to stop his men from driving sheep across the Rangitata River between mid-afternoon and early the following morning because of the presumed risk of higher flow rates at that time.[52]

During a formal debate in the Canterbury Provincial Council on an engineer's report about flooding in the Waimakariri River, Mr Beswick disputed some of the data and the inferences drawn on the grounds that the latter were contradicted by Māori tradition, to which the editor of *The Press* on 26 September 1864 tartly responded, 'we can hardly agree with an intelligent member of the Provincial Council who stated . . . that Mr Doyne's [surveyed] levels were contradicted by Maori traditions'. Later that year the 7 November issue of the *Grey River Argus* reported that Greymouth had experienced the heaviest flood in the memory of the oldest European resident and 'higher than can be remembered by any of the Maoris who have been settled here for many years'. The report in the *The Press* on 6 February 1868, however, indicated that even though Māori might have recalled major environmental events they had not always correctly forecast their reccurrence.

PLANTS AND ANIMALS

Māori had a close knowledge of native plants and animals, including where they could be found. In 1838, John Watson told a committee of the House of Lords in London, 'They assisted me in directing my attention to plants and flowers; where they thought there was a particular plant I had not seen, they would bring it, expecting some little remuneration; tobacco, for instance.'[53] Baron Hugel, who collected plants in the North Island during the 1840s, reiterated what missionaries had earlier reported, that 'there is not, in the northern island at least, a single tree, vegetable, or even weed, a fish, or a bird, for which the natives have not a name; and that those names are universally known'.[54] Another early collector was the German botanist, Dr Ernst Dieffenbach, who, according to Canon James Stack's Māori informants, was 'just a collector of rubbish. He went everywhere collecting leaves and flowers and roots and stones, and actually paid people to carry these worthless things from the interior to the coast. What could possess the man to act so foolishly?'[55] During his travels, Shortland was told by local Māori that kahikatea of excellent quality grew beside a river on Stewart Island, and in his report on the Canterbury Block, published in the *Lyttelton Times* on 8 March 1851, Frederick Weld wrote, 'The old natives told me that kakapos were numerous in the inland black birch woods; I did not find any. The kiwi, the natives said, was rare but sometimes found near the sources of the Awatere, where one night I imagined that I recognised its cry. Wood hens [weka], blue, whistling and paradise ducks formed our chief food in the latter part of our expedition, and indeed we had little else to eat; they were all very tame: as to the former, their coolness was often provoking.'

In her privately printed reminiscences, Mrs Michael Studholme, the first European woman in the Waimate district, south Canterbury, recorded the following Māori names for plants and animals: 'inini [*sic*], kaka, konini, manuka, matipo, ohau [possibly *Pseudopanax* sp.], raupo, totara and tutu'.[56] And that ever-interested and observant settler Lady Mary Barker recorded Māori and European names for the plants and animals on the property that she and her husband occupied in the upper Selwyn Valley. She was particularly assiduous in seeking out information about the new land, and while on a visit to neighbouring Rockwood Station feared 'I must have wearied our dear, charming host by my incessant questions about the names of the trees and shrubs, and of the habits and ways of the thousands of birds. It was all so new and so delightful to me.'[57] European settlers heard or read about Māori traditional knowledge of environmental change, with stock drover Edward Chudleigh recording in his diary: 'You find a good deal of burned wood all over the hills [in the Otago gold-fields]. The Maori have two traditions, one is that the northern tribes burned it when at war with them, the other is that they burned the country to get rid of the moa bird whose bones have been found in vast quantities here . . . . the whole country was wooded once, that is [now] plain.'[58] In another opinion, T. D. Burnett reported fossil evidence of a primeval forest in the Mackenzie Country about which he believed Māori had no tradition.[59]

Māori biological knowledge could be sophisticated, as the following item from the 6 September 1865 issue of *The Press* showed:

> HOW TO PUT POSTS IN A FENCE: it is a universal practice of the Maoris in put-ting posts into the ground for a post-and-rail fence, to put the post upside down, that is with the end which stood highest in nature into the ground. They assert that posts so fixed will last a much longer time than posts put in in the usual manner. As the aboriginal races are usually right in all matters which spring from a minute observation of nature, those who are not too proud to learn from Maoris may imitate this custom with advantage.

This practice may also inhibit sprouting from the exposed end of the post.

MĀORI AS CULTIVATORS

Reports by explorers and early settlers reveal that not only were Māori inno-vative horticulturalists, but their farming practice also showed a concern for sustainability. Elsdon Best found that Māori cultivators recognised approxi-mately 50 different soil types, appreciated free-draining and friable soils on river terraces, knew how to lighten stiff clays with gravel and understood the value of minimal cultivation.[60] Their gardens were neatly set out, largely weed-free, and typically used for three to five years before being abandoned.

Whereas Māori relied on a few staples that they harvested or grew in suitable areas, they also transplanted some food species before the colonial period and adopted new sources of sustenance in the early years of contact. Wherever they could be reliably grown, kūmara were the subsistence food but they were not the only plants cultivated at this time. Fearon referred to four karaka trees – a coastal species rarely seen far inland – that had been planted midway between the Wairarapa coast and the pā at Gladstone.[61] When the fruit there were ripe, Māori used to trek down to the coast to harvest fruit from the extensive natural stands of karaka. Māori also planted the terminal leafy shoots of tī – thereby establishing or enlarging groves for later harvest – as well as off-shoots of harakeke.[62]

On 29 May 1773, Captain James Cook recorded in his journal that 'turnips, carrots and parsnips ... together with the potatoes, will be of more use to them [Māori] than all the other articles we had planted. It was easy to give them an idea of these roots by comparing them with such they knew.'[63] Bawden reported that, with Samuel Marsden's assistance, Ruatara had introduced wheat to Northland. Iwi were surprised, however, when the fully grown plants lacked edible tubers. When he was given some grains of wheat, Houpa asked Marsden how long he would have to wait between planting and harvesting. In his journal Cruise reported seeing a two-month-old plant of the edible garden pea growing under cultivation in Northland. The seed had come from Coromandel and the developing plant was tapu, which meant that it would not be harvested until it was ripe to be eaten or to be kept as seed. Elsewhere in Northland, Cruise found that Māori 'eagerly adopted the improvements we pointed out to them in their system of agriculture; and they were very grateful for the European seeds distributed among them. Many chiefs had very fine crops of peas before we sailed, which they promised not to consume, but to save the seed, and sow it again: the water melons were in great luxuriance; and the degenerated cabbage and other vegetables were much improved by their being taught how to transplant them.'[64] The officers of the *Dromedary* left behind seeds of watermelon, peas, oak and orange for cultivation.

Samuel Marsden was impressed by the excellent crops of potatoes – which were even cultivated by Māori living along the southern coasts of the South Island, far from where kūmara could be grown, for sale to sealers and whalers – but expressed concern over the prospect of soil nutrient depletion if too many consecutive crops were taken from the one garden. According to one settler, the standard of Māori cultivation was low: 'they take a potatoe planting stick, and tread it into the earth, pull it out then in with the potatoe, and poke them out with a stick in the same manner. The only preparation is to burn the fern and sticks, and put in the crop, and they do no more until they dig the potatoes.'[65]

Other contemporaries, however, complimented Māori on their methods for cultivating maize, sweet potato and white potato. So proficient were Māori cultivators of recently introduced root and leaf vegetables that early European settlers in Christchurch[66] and Napier were sustained by their harvests. The 24 June 1868 edition of the *Otago Daily Times* reported a piece first published in the 26 May 1868 issue of the *Hawke's Bay Herald* about a gift of fifteen tons of potatoes from Chief Tareha to Napier residents.

By 1848, Ngāi Tahu at Arowhenua in south Canterbury were growing wheat and potatoes, pumpkins, maize and vegetable marrows while at the same time gathering traditional foods from mahika kai.[67] An article published in the *Nelson Examiner* on 29 March 1845 described a Māori settlement on the site of Picton covering about 20 hectares. In part of the settlement, potatoes, cabbages, turnips, Indian corn, kūmara, melons and pumpkins had been planted. Farther south, potatoes, maize, wheat, melons, cabbages and other introduced vegetables were grown by Māori for sale to settlers, but kāuru, sea fish, eels and whitebait remained staple food items for tangata whenua. The Māori propensity for cultivation was bolstered by their wish to trade: for their own food they relied on a combination of planting, hunting and gathering, and combined traditional with new forms of sustenance. For their part, European settlers in southern New Zealand showed little appreciation of Māori food and instead grew or brought in their staples.[68]

COMMUNICATIONS BETWEEN MĀORI AND PĀKEHĀ

It is impossible to prevent colonization; but it will be colonization of the worst kind which must annihilate the [Māori] people. The generality of the present European population, now residing in New Zealand, will destroy, will extirpate and annihilate the [Māori] people; it cannot be otherwise. Many of those men are superior to the missionaries in their influence. A native looks to the people who will give him most payment. What is a man who understands Greek or Latin, or drawing, or music, or has superior manners? The native does not like him so well; but those who come nearest to themselves will have most influence.[69]

In a submission to the House of Lords, the Reverend Mr Frederick Wilkinson described Māori as 'particularly scrupulous in not infringing on another's property. For instance, in returning from Waimate [in the Bay of Islands] I was going to take some peaches from a tree that was there [but] the native that was with me told me I must not do so. When I had gone a little way further he allowed me to take [fruit] from another peach tree belonging to a relation of his. It appeared that he could take that liberty with a relation'.[70] The following decade, Shortland observed that Māori 'would generally only come in contact with the European population when they found it in their own interest to do so, and then as guests, and not neighbours – a mutual advantage – because the natives being tillers of the soil and the Europeans farmers at first principally stock-keepers, their proximity to each other gives rise to frequent disputes about the trespass of cattle, and remuneration to be paid for damage done'.[71] In the same vein, Gillespie noted that 'One night Mantell overheard the Maoris arranging amongst themselves to give him false names the following day when he would be investigating land and setting aside the reserves.'[72] Another administrator, Henry Sewell, described Māori at Rapaki Pā on Banks Peninsula as

> civil and harmless, I cannot find anyone who can give an account of them.
> In truth they are left to themselves, to get civilised if they can and how they can
> by contagion. Nothing is being done for them till the Colonists themselves have
> the full management of matters. Nobody is responsible at present. Sir George
> Grey and the Bishop [Selwyn] used to come now and then, and look at them, and
> talk about Schools and so forth, and then go away, and Sir George would write
> a flashy dispatch about himself and his own virtues, making the English world
> believe that it was all owing to him that the Natives were not dining and supping
> off the Settlers, all of which is detestable humbug.[73]

Until the early 1860s Pākehā supplemented and extended their own observations with information they had obtained directly or indirectly from Māori about weather and climate, hydrology, living things, topography and mountain passes. In time they became less reliant on indigenous environmental knowledge and followed their own procedures for making observations, then recorded, analysed and communicated information about the environment. In doing so, they were acting like the occupants of a land that lacked an endemic population and was without a history of observation, consideration or analysis of environmental features and forces.

CHAPTER TWO

Settlers Learning about Wind, Warmth and Rain

A good deal of nonsense has been written about the climate of New Zealand. We have heard of the balmy air, the beautiful blue of the sky, the perpetual greenness of the landscape, and other equally pleasant elements, found only in novels and fairy tales. Not a few who have gone to the colony have been grievously disappointed at finding that it can blow as hard, and rain as heavily, there as in less favoured countries. This sort of over-colouring is very foolish as well as very wrong.[1]

The first generation of European settlers had received little reliable information about the weather in southern New Zealand, but had high expectations of life in the new land. They were, however, fortunate in taking up residence at a time when communications between distant places were improving, science and engineering were developing rapidly, and people in the British Isles were seeing the beginnings of weather forecasting based on several years' numerical observations from a network of well-equipped weather stations. For at least two decades, pioneer settlers had few reliable accounts of environmental conditions to guide them, and the geographical position of New Zealand in the centre of the ocean hemisphere rendered northern hemisphere models and comparisons unhelpful. Despite the difficulties, successful European settlers learned about the normal climatic conditions of southern New Zealand and became adept at discerning and responding to signals of a change in the weather. This was not merely a hobby interest of people with time on their hands but an essential skill for economic well-being. Like many skills, it had to be learned on the job, supplemented by observations made in the surrounding area as well as useful information heard at gatherings and from visitors, or read in newspapers and magazines.

This chapter will show how settlers began to learn about the weather and climate of southern New Zealand and then to diversify their observations and data collection. Initially, they took note of weather conditions on their own properties, recording them, discerning patterns and making simple observations that they then compared with the experiences of people elsewhere. This in turn allowed settlers to identify and interpret the signals of normal weather events and then to make informal weather forecasts. As their knowledge of normal weather conditions, as well as the consequences for livestock and pastures, deepened, so did settlers note variations across the local area, the region and, to a lesser extent, the country. In addition, while those comparisons were initially geographical, they soon included a temporal component, which allowed some sense of the length and character of seasons to develop. Within a decade or so they were able to compare their recent experiences of the weather with reports or recollections of conditions in earlier times. This gave them a sense of the timing and characteristics of seasonal weather for their properties, and how conditions could change from one year to the next. Limitations in measurement technology and communications, and the early emphasis of meteorology on statistical summaries of numerical measurements, rather than scientific explanations rooted in mathematical principles and physical laws, made reliable forecasts of unusual weather events – such as prolonged spells of snow, unseasonal outbreaks of very cold air, heavy rain leading to widespread flooding and extended periods of significantly lower-than-normal rainfall – almost impossible to achieve, but that did not stop some intrepid souls from trying.

ONE CHALLENGE AFTER ANOTHER

With assured financial backing, a nineteenth-century farm or station might survive several months of low market prices for meat, wool and hides, but the impact of a spell of bad weather was immediate and long lasting. Little wonder, then, that the state of the weather was usually the first written entry – on some days the only words – in a farm or station diary. A particular weather system could be beneficial in one season and a major setback in another. Winter snow, for example, normally resulted in steady inflows of melt water into the top-soil, supporting a spring surge of plant growth. Snow only became a threat if it persisted and lay deep enough to trap sheep for several days. A snowstorm, a hard frost or a spell of cold, squally weather in early summer, however, could kill tender shoots and reduce crop yields, as Joseph Davidson discovered when Southland experienced several outbreaks of very cold air during the summer of 1893–94. On 5 December 1893 he wrote in his diary, 'The day was very stormy … I hear Mr Gillanders lost a lot of his [recently] shorn hoggets – with the cold blast they are lying dead everywhere [in the district].' Ten days later he recorded that heavy frost had severely damaged his potato crop, and at the end of January he described a day of very low barometric pressure and stormy weather:

> The stormiest day we have had since winter. There was a very heavy thunder-storm, hail, wind and rain, with the temperatures very low – the cold was phenomenal for this time of the year, the hills and mountain tops got a heavy coating of snow – If it should come on a warm rain there would be a heavy flood – It is to be hoped it [the snow] will come away gradually. I see by the [Otago] Witness a vessel has arrived [in port] reporting there are large icebergs due west of Stewart Island …. [Ice in the] Antarctic region must have broken loose, for there has been nothing but reports of ice bergs. Ships getting hemmed in by them, some [ships] coming into collision with them, and no doubt many [ships] have gone down by coming into contact with them – we will never hear of [them again].

During summer, several weather systems had a significant impact on the work routines of farm and station. Strong dry winds could bring about a delay in sowing turnip seed for winter feed and spreading manure or guano, raise dust from ploughed fields and dry riverbeds, flatten ripening grain, leave the soil too hard for a fencing gang to dig post holes or drive in standards, and cause heat stress in shearers. Even a light drizzle during shearing could force abandonment of work for a day or longer because damp wool makes the animal too heavy to manipulate on the shearing stand and can be a health hazard for the shearer. For that reason, even recently established pastoral properties usually had sheds large enough to hold several hundred sheep under cover in readiness for shearing the following day. Persistent rain in December–January, however, was of greater concern. In the words of the manager of Ida Valley Station, written to the absentee landholder on 2 January 1877, 'The [rainy] weather is causing great delay and trouble everywhere down country. I believe at Rickland's Station only 8000 ewes were shorn in nearly three weeks.' Rain in late summer also delayed haymaking and could reduce its quality.

Blown dust was a frequent occurrence throughout southern New Zealand, and not only in rural areas after tussocks had been cleared and the soil ploughed. In January 1849 the surveyor Charles Torlesse wrote in his journal about cold northeast winds picking up sand grains from the dunes on the seaward flanks of the Heathcote Estuary and transporting them inland, as well as being 'smothered with sand' raised from the dry bed of the Waimakariri River during a northwest gale.[2] Two years later Edward Ward recorded in his diary that 'At 3 a.m. on 5 February the wind rose in gusts, and one filled the [Lyttelton Assembly] rooms with such a cloud of dust as perhaps was never seen before in a ballroom.'[3] On 6 February 1853, soon after he arrived in Christchurch, the acerbic Henry Sewell wrote in his diary that 'When it blows (according to all accounts one may ask when it does not blow?) the inhabitants must be smothered with dust but they make no complaint on this score.'[4]

Strong winds in autumn and early winter could delay harvesting and threshing, but between late autumn and early spring a greater concern for a landholder was the combination of a heavy fall of snow followed by a partial thaw and several days of severe frost restricting access to feed by sheep and cattle. Frozen soil also made it difficult for teams of draught horses to plough arable land in preparation for sowing grain and root crops during spring, as Joseph Davidson recorded in his diary on 22 August 1878: 'Alek ploughed in forenoon [for the] first time since 30 July owing to snow and storms – very little ploughing [has] been done anywhere [in the Waikaia area].' Persistent low cloud and heavy rain in early winter, especially from the easterly quarter, delayed mustering in the hill country and caused intense frustration for landholders, as is evident in

Storm-tossed cabbage trees on the shore of Lake Wanaka, early twentieth century.
GEORGE CHANCE FRPS: HOCKEN COLLECTIONS, UARE TAOKA O HĀKENA, UNIVERSITY OF OTAGO

the following diary entries by members of the McMaster family at Tokarahi in north Otago over a period of six days in late winter 1886:

> Very wet weather. It looks like it isn't ever going to stop raining. [It had been raining most of the previous seven days.] Wind SE and rain from there. (16 August)
> Wet, wet, wet, wet, wet, wet, etc, etc, etc, etc, etc, etc. Raining still [in evening]. (17 August)
> Raining. When will it stop. Somewhat harder than yesterday. Wind NE, SW, NW, SW, etc, etc. Evening raining still. (18 August)
> Weather rather better, but far from good. (19 August)
> Very fine day. Sun and heat. (20 August)
> Wet, wet, wet. Wind NE. Evening fine. (21 August)

Farms and stations were particularly vulnerable to abnormal weather conditions in spring, as the manager of Ida Valley Station informed his landholder:

Lambing has started (1 October [1893]) . . . . we just escaped the rough days at end of September. Since then we have had perfect weather, and it looks settled now. After the rain we have had, and with weather like this, we should soon have grass, though there is mighty little . . . . I never saw the sheep so poor here as they have been for the last 6 months, and I sincerely hope they may never be so bad again as long as you or I have anything to do with Ida Valley. The winter was no doubt bad, but it was want of late summer and autumn rain that did the mischief! Given a fair amount of rain then, and consequently grass on our lower country and sheep in [good] condition, with our fencing and exposure we can stand almost any winter.

Strong northwest winds in spring could force postponement of any remaining cultivation, but a stiff southwest wind accompanied by light showers was beneficial in keeping down the dust while ploughed land was being harrowed in preparation for sowing pasture plants, root and grain crops. Across southern New Zealand another concern for residents of farms and stations alike was a spell of cold wet weather from the easterly quarter during lambing. On 17 September 1883 McMaster recorded, 'Showers morning from NE and SE. Damn the NE and SE. A week of this weather at lambing time is quite enough.' Abnormal weather in late spring had a flow-on effect. On 4 December 1852 the *Otago Witness* reported that November had been unusually wet and cold in Otago, with the mean monthly temperature 6° Fahrenheit lower than in November 1851 and 5° lower than for the same month in 1850. Plant growth was consequentially less than expected and the shearing delayed.

WEATHER AND CLIMATE IN SOUTHERN NEW ZEALAND:
PERCEIVED AND EXPERIENCED

Settlers tended to arrive in New Zealand with preconceptions about the weather that were based on information provided before they left Britain and from loosely conceived geographical comparisons and hearsay. First of all, there was an expectation that New Zealand's climate might be understood in

Rangitata River

View across the Canterbury Plains from the intake of the Rangitata stock watering race, late nineteenth century. HISTORIC PHOTO ALBUMS, GENERAL SOUTH CANTERBURY THEMES, S.C.4, SOUTH CANTERBURY MUSEUM, TIMARU

relation to that of the British Isles, and secondly, by comparing northern and southern hemisphere latitudes, that the closest comparison in fact would be with the Mediterranean climates of northern Italy and its environs.

In the handbooks published in England for their guidance, intending settlers could read about the economic, environmental, governance, political and social conditions they should expect to encounter in the new land, as well as how they compared with those of the British Isles. James Preston kept a clipping from the *Mount Ida Chronicle*, drawn from a *Melbourne Age* piece reproduced in the 6 April 1872 issue of the *Otago Witness* about 'a very handy and well got-up book of hints to emigrants [that] has been prepared by Mr Hall, the Clerk of the Legislative Assembly of Tasmania. Mr Hall's book has been sent to every newspaper in Great Britain.' Even sketchy accounts of seasonal variations in rainfall and temperature were of interest to individuals intending to become arable farmers or to raise sheep and cattle in the new land. It was not until the first generation of European settlers had experienced life in New Zealand that they realised climatic comparisons with southern Europe were flawed.

Because the countries situated on the diametrically opposite side of the globe – New Zealand's antipodes – extend from north Africa, across the Iberian Peninsula and into the Bay of Biscay, it was assumed that weather patterns between the two land masses would be similar. Pamphleteers such as Charles Hursthouse compared New Zealand's weather with that of the Mediterranean basin,[5] and despite the growing numbers of settlers experiencing the weather in New Zealand as it was, belief in a southern hemisphere Tuscany persisted for some time.

It was observation and record-keeping that dispelled these northern hemisphere assumptions. Reliable knowledge of the country's diverse physical environments, as well as how they varied over short and long periods of time, came with personal experience in the new land. The weather was not often balmy and in their diaries and letter books settlers sometimes wrote with disappointment about the cool temperatures and the persistence of rain. The settler who longed for the classic Mediterranean climate, in which the bulk of the year's precipitation falls in winter and the summer is hot and dry, was soon disappointed. Edward Ward's experiences in Canterbury during the first half of 1851 indicate the extent to which early settlers in the south were buffeted by the weather:

> Cool sunny day of real New Zealand pleasant weather. (10 February)
> A lovely real New Zealand day, warm sun and cool breeze and delicious haze over everything. (11 February)
> As usual with newly arrived [immigrant] ships, New Zealand put on its most forbidding aspect gloomy and misty, ending with a really wet evening, a downpour the worst that there has been since we came. (1 March)
> Last night had a fresh frosty cold about it [and] this morning had the same feeling, something of a bracing English autumn, and very pleasant. (19 March)
> Many consider this is the hottest day we have had since landing. I almost think so and yet we ought to expect English October weather . . . . When is this long, long summer to end? (27 March)

Today the same extraordinary changes of temperature occurred which have given character to this autumn. The morning was showery and cold; forenoon sunny, calm and warm; afternoon blowing hard with rain from S.W. and bitter cold. A hail storm visited us today, and tonight it feels quite frosty. (6 April)

A real May day, finer, I warrant, than they have in England. (1 May)

Surely finer weather never was seen at any time of the year in any country. (5 May)

Tonight it blows raw and gusty from a quarter too near N.W. to be pleasant; however we should not complain if this pleasant weather comes to an end, it has cheered us so long. (8 May)

Even at nine o'clock in the morning it was as warm as a summer's day in England and continued fine all day. Nothing can surpass the beauty of this winter weather. (26 May)

The shortest day and almost the loveliest I have yet seen in New Zealand. (22 June)[6]

Two years later, Henry Sewell recorded comparably mixed experiences of the weather in mid-Canterbury:

A cold wet afternoon by no means answering to the idea of an Italian climate. (10 February)

Oh this New Zealand climate . . . this south of France with Italian skies. (8 March) [Sewell wrote these ironic words during a brief visit to Wellington.]

Such a day towards the end of autumn answers to November in England and reminds one of the difference in latitude. The general temperature is undoubtedly higher and better than England. (4 May)

Yesterday and today are lovely . . . just such weather as one would desire for new-comers, like warm spring days, an English May, whereas this month answers to November in England. (7 May)

The weather [is] detestable. Rain in all its forms from mild, gentle, insinuating mist, to fierce tropical torrents, with all the intermediate varieties. Tuesday night it poured, water spouts, the gullies all run in cataracts, slips of land in all directions, and the roads impassably quagged. (5 July)

Of all the mutabilities, the New Zealand climate is the most mutable. (29 October)

Barring this episode [a northwest gale], the weather has been on the whole beautiful, and I must withdraw my maledictions against the clime. It certainly does answer pretty well to the descriptions one used to hear of it, when people were painting up the Colony. (28 January 1855)[7]

In their diaries, the residents of The Point Station – which extended from the foothills of the Rockwood Range to the northern bank of the Rakaia River, 5 kilometres west of Windwhistle – did not compare the area's seasonal weather with that of northern Italy or the British Isles where they were born.[8] The principal writer did, however, share Sewell's sentiments about the weather in Canterbury and often expressed his opinions in comparably ironic terms: 'Howling s'wester, fearful snow. A devil of a day . . . . Eight inches of snow at night. Verily this is a cursed country' (25 August 1868); 'NW gale, thunder and heavy rain . . . . One of the cursed days peculiar to this great country' (15 October 1868); and 'SW and heavy snow in morning; strong NW afternoon. Simply a charming day' (28 September 1869).

The first generation of European settlers expected New Zealand to have a rainy season. The land surveyor John Barnicoat, who was employed by the New Zealand Company and based in Nelson, recorded in his diary on 19 September 1842: 'A considerable amount of rain; this is evidently the rainy season of New Zealand.' On 4 November 1842 he noted, 'We hope to see symptoms of a gradual setting in of the dry season as the rain for a week past has not been sufficient to keep the streams and marshes up to the point we found them on arrival.'[9] A decade later Sewell also anticipated a rainy season in Canterbury, and on 5 July 1853 wrote, 'But this is the rainy season, so they say; the New Zealand season is always [rainy] so far as our present experience goes.'[10]

The handbooks available to settlers seldom hinted at environmental differences between the North and South Islands and did not represent the great environmental diversity of New Zealand, where weather systems can vary between places a few tens of kilometres apart and in the course of a day. A deeper understanding of local weather systems, including their geographical and historical variability, is evident in the impressions of Frederick Teschemaker, written in mid-July 1857 while he was travelling by horseback through the eastern South Island from Christchurch to Dunedin, thence farther south:

Most of the country from Ch[rist]ch[urch] to Dunedin (with the exception of the last 35 miles) I like very much, especially the country about the Waitaki [River], more particularly the south side in which Wm Teschemaker's run is. The climate is milder. Nice undulating country, pretty little valleys and streams of water, but unfortunately with almost as great a want of wood as in Canterbury. They get, however, none of those strong SW winds that we are so troubled with [and], from what I hear, not so wet as Chch. I expect that after Nelson, the Waitaki climate is the finest in this Island .... The country below Dunedin is not so good, [and] the flat parts are wet and full of creeks. The climate is undoubtedly very inferior, there is such a prevalence of SW winds and there are no high mountains like in Canterbury to break them and draw the rain away, the breezes come off the Pacific un-interruptedly and favour country with more moisture than is agreeable. In fact it is very wet and cold both overhead and underfoot.[11]

The new settlers gradually amassed sufficient local experience to recognise that atmospheric cells of high or low barometric pressure, disturbances embedded in the flow of the circum-Antarctic westerlies, occasional cyclonic systems of sub-tropical origin and outbreaks of very cold sub-Antarctic air were the principal weather drivers: locally, regionally and nationally, as well as daily, seasonally, and for a year or longer. In the process, they began to understand the weather of their adopted land as it actually was, rather than in comparison with other places with an entirely different suite of geographical conditions.

WEATHER COMPARISONS

When a recently arrived settler wished to put into context a major winter snowfall in mid-Canterbury, the only comparison he or she could validly make was to a similar event previously experienced, heard or read about in Great Britain. Clearly, a settler could not compare a spell of bad weather with earlier such events on his New Zealand property, and the chances are that his neighbours were as unaware of them as he was. With sufficient experience on the land, supported by statements in the press and the recollections of neighbours, a settler no longer had to draw comparisons between mid-Canterbury and British weather, but could find context in the range of local weather conditions experienced first hand or by someone else. In order to form a picture of the growing ability of settlers to compare weather conditions, I compiled comparative statements in farm and station diaries about individual weather events and associated environmental conditions on the property, as well as the writer's observations of, or reports about, the situation in the district, the province and the nation. The findings are depicted in Figure 2.1. The horizontal axis for each diagram is distance in kilometres between the property and the compared place, and it is shown on a logarithmic scale because we tend to know more about, and have a greater interest in, what happens nearby. The vertical axis concerns each comparison recorded in a diary or letter book, and whether it was inferred, experienced or measured. Depending on the circumstances, rural people employed a variety of measurement scales when they observed and recorded environmental conditions and effects.[12] The weakest of these, known as the nominal scale, is where words or symbols are used to classify objects: for example, 'rain' or 'wind'. The ordinal scale is stronger, and indicates a ranked relationship between successive classes: for example, 'drizzle', 'light rain' and 'deluge'. Next is the interval scale, where the span between two consecutive records is known but the units of measurement are arbitrary and there is not a true zero: for example, 'as much rain as yesterday' and 'twice as much rain as this time last year'. The ratio scale is the strongest because the interval between two consecutive values has been measured and there is a true zero: '15 inches of rain captured in a rain gauge over the past 24 hours' and 'an air temperature of 28 degrees Centigrade in a shaded site at noon' are examples of this.

Each of the diagrams in Figure 2.1 summarises comparisons made by eight European settlers during the five decades from the late 1840s to the late 1890s, and the figure shows four trends: from conclusions based on beliefs and assumptions to conclusions stemming from local observations; to greater reliance on scientifically based observation; to the use of more precise record-keeping and orderly information-gathering for statistical analyses; then comparisons across time and space.

Early in the period of organised settlement – before environmental observations made by European settlers could be collated, published and compared – individual settlers sought context for their local experience in what they knew or believed about environmental conditions in Great Britain and Mediterranean Europe. Initially, most settlers' geographical and historical comparisons related to temperature, less frequently to rainfall, and rarely to strong winds, floods or snowstorms. Where the comparison was with Great Britain, it was implicit that the writer had directly experienced the named environmental condition before shifting to New Zealand. Where the comparison was with the south of France or Italy, however, the writer tended to rely on

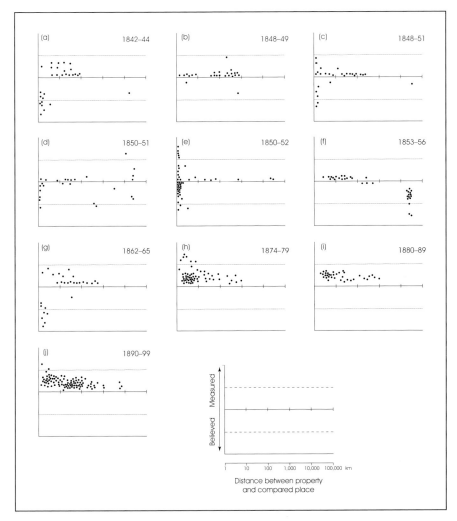

FIGURE 2.1 A pictorial representation of weather comparisons collated from entries in ten runs of settlers' diaries, and arranged chronologically: (a) John Barnicoat in Nelson Province (1842–44); (b) Thomas Ferens of Waikouaiti in coastal east Otago (1848–49); (c) Charles Torlesse in mid-Canterbury (1848–51); (d) Edward Ward on Banks Peninsula and environs (1850–51); (e) Francis Pillans at Inch Clutha in south Otago (1850–52); (f) Henry Sewell of Christchurch and environs (1853–56); (g) Edward Chudleigh in mid-Canterbury (1862–65); and (h, i & j) Joseph Davidson at Waikaia in northern Southland (1874–79, 1880–89 and 1890–99, respectively). SOURCE: INFORMATION COLLATED FROM THE NAMED SETTLERS' DIARIES AND LETTER BOOKS (a, b, e, h, i & j): MALING, 1958 (c); MCINTYRE, 1980 (f), RICHARDS, 1950 (g); AND WARD, 1951 (d).

perceptions and memories of things read. Amongst well-established settlers, comparisons between local and northern hemisphere weather systems, and their associated environmental and economic impacts, had virtually ceased by the 1860s. Even detailed comparisons between New Zealand and Australia were few. One example of that comes from the diaries of Joseph Davidson, who was born in Scotland, travelled to New Zealand via Australia while still a young man, and settled in the Switzers / Waikaia area of northern Southland in 1863. Some summers, he noted hazy skies and falling ash, and explained these as the consequence of bush fires in southeast Australia. Nowhere in his diaries, however, did he compare the weather systems and associated environmental conditions that he was experiencing in northern Southland with what he had known as a child in Scotland. Comparisons between southern New Zealand and Western Europe helped newly arrived settlers come to appreciate that these two areas have different climates and distinctive weather systems. Later, any comparison between the two was in relation to marketing wool and other primary products in Great Britain.

The second trend was a progressive shift from perception to close observation of weather conditions and their associated impact. This showed in how settlers recorded their observations and drew comparisons: for example, from 'snow on the hills' to 'snow to low levels on the hills and extending onto the plains'; and from 'wind in the mountains' to 'strong northwest winds in the mountains'. The third trend was towards more precise measurement and systematic observation on the property as well as across the district, and the fourth was to more frequent geographical and historical comparisons, often involving mathematical averages. By the end of the nineteenth century, landholders were recording a range of daily weather conditions and environmental events on their properties, and making greater use of measuring instruments.

WEATHER PATTERNS

John Wither observed each day's weather almost without break from September 1864 until the end of the nineteenth century and recorded those observations in his diaries. Wither was employed on a pastoral property near Otematata in the Waitaki Valley in 1864 before taking up the lease on Moonlight several kilometres north of Arrowtown three years later. In 1870 he and his young family shifted to Sunnyside Station on the southwest shore of Lake Wakatipu. Each day thereafter he briefly described cloud cover, precipitation, temperature and wind, but seldom recorded barometric pressure, air temperature, precipitation or wind direction exactly. I was able to collate and statistically analyse his descriptions of daily weather conditions because he regularly used a small number of descriptive terms for precipitation, temperature and windiness. These allowed me to cluster his terms and arrange them in order of increasing magnitude. The first class – 'mist', 'fog' and 'dew' – represented the least amount of precipitation. 'Showers' and 'showery' were bursts of light rain; 'wet', 'rain' and 'rainy' were longer spells of rain; and 'very wet' and 'heavy rain' made up the fourth class. His occasional references to 'storms', 'stormy weather' and 'thunderstorms' formed the fifth class, and 'snow' the sixth because slow melting usually recharged stores of soil moisture, something that the Scottish manager of Ida Valley Station, and Wither's contemporary, also understood: 'What this [dry] country wants is the next two months under snow'; and 'On the last day of the month [July] from 2 to 3 inches of snow fell all over [the run]. It has now [4 August 1890] almost disappeared without affecting the creeks, and moisture now in the ground will give us a start out [with spring feed].'

I was also able to cluster Wither's references to air temperature. The first class was 'frozen', a word he seldom used. 'Hard frost' and 'very hard frost' formed the second class, with 'frost' and 'frosty' the third. Wither used 'cold' for what felt to him like very cool but not frosty conditions (the fourth class). The fifth class contained all of his references to 'mild' conditions and 'thaw'. 'Warm' was the sixth class and 'very warm' the seventh, while 'hot' and 'very hot' formed the eighth class.

I then assigned each of his references to wind velocity to the most appropriate of five ranked classes, ranging from 'calm' and 'breeze' in the first, through 'blowy' in the second, to 'very blowy' in the third, 'squally' in the fourth and 'rough' in the fifth.

Those descriptors were treated as interval data, and the graph of annual averages (Figure 2.2) shows 'blowy' to 'very blowy' conditions prevailing from 1867 to 1890. In contrast, average temperatures were mild in 1872, 1880–81 and 1885–87, but cooler in 1868, 1877, 1884 and 1888. Average annual precipitation peaked in 1867–68, 1875–78, 1884 and 1888, was low in 1872–73 as well as from 1880 to 1883, and very low in 1886. There was a shift from wetter weather in the late 1860s to drier conditions in the 1880s. These trends were even more

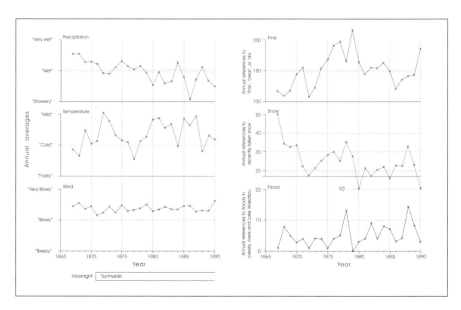

FIGURE 2.2 Average annual precipitation, temperature, windiness, incidence of 'fine' days, fresh snowfalls and days when rivers and streams were in flood on John Wither's sheep stations at Moonlight, near Arrowtown (1867–70), and Sunnyside, across Lake Wakatipu from Queenstown (1870–90).
SOURCE: INFORMATION EXTRACTED FROM JOHN WITHER'S DIARIES: SEE APPENDIX NOTE 8 FOR DETAILS

apparent in the annual occurrences of 'fine' days, 'fresh snow' and 'floods'. 'Fine' days were relatively frequent in 1871, 1879 and 1890, and snowfalls were more common in 1867, 1878 and 1888. The incidence of flooding scarcely changed between 1867 and 1877, peaked in 1878, and rose from no reported floods in the mostly fine warm year of 1879 to reach a peak in 1888. The variable weather recorded by Wither on his property north of Arrowtown between 1867 and 1870, and on the flanks of Lake Wakatipu from late 1870 to 1890, suggests that he was experiencing swings between La Niña and El Niño conditions.

REGIONAL WEATHER

More than 10,000 of the total number of dated entries in the nineteenth-century diaries and letter books that I read referred to wind, water or temperature, all of which showed considerable variability across southern New Zealand. Settlers were primarily interested in the weather and climate of their respective home

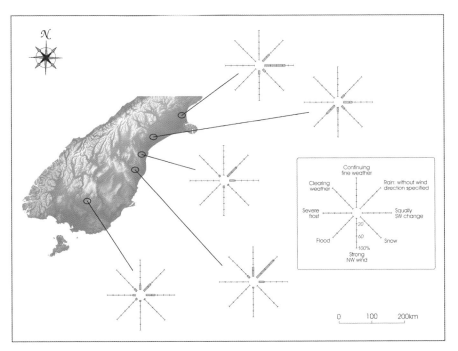

FIGURE 2.3 Relative number of references to each of eight environmental conditions noted in five settlers' diaries: Christchurch and environs (Munnings, 1859–66), mid-Canterbury (Chudleigh, 1862–65), south Canterbury (Gunn, 1918–35), inland north Otago (McMaster, 1885–93) and northern Southland (Davidson, 1874–1900). SOURCE: INFORMATION EXTRACTED FROM THE NAMED SETTLERS' DIARIES; SEE APPENDIX NOTE 8 FOR DETAILS

areas (Figure 2.3), and the diagram summarises weather observations recorded in five sets of diaries: Joseph Munnings for mid- and north Canterbury as well as Christchurch and its environs; Edward Chudleigh, mostly for lowland Canterbury;[13] John Gunn for the south Canterbury downlands northwest of Waimate;[14] the McMaster family for their property at Tokarahi on the southern flanks of the Waitaki Valley in north Otago; and Joseph Davidson for his farm at Switzers / Waikaia in northern Southland. Those five runs of diaries were chosen because they contained detailed qualitative accounts of daily weather conditions over periods of four or more years, allowing me to collate references to eight distinctive weather conditions: (1) continuing fine weather, (2) rain from an unspecified direction, (3) a squally southwesterly change, (4) snow, (5) strong northwest winds, (6) heavy rain and flooding, (7) severe frost, and (8) clearing weather.

Joseph Munnings, whose work in the Christchurch area and mid-Canterbury often kept him out of doors for days at a time, was especially sensitive to squally southwest changes. For stock drover Edward Chudleigh heavy rain and flooding in streams and rivers were of greater moment and predominated in his accounts of environmental conditions on the plains and foothills of mid- and south Canterbury. They were followed by changeable westerly weather at any time of the year and snow in winter. In south Canterbury the mixed crop and livestock farmer John Gunn mostly referred to rainy weather, and in north Otago the McMaster family did much the same. Joseph Davidson's weather observations on his farm in northern Southland showed almost equal representation of each of the eight classes, reflecting that area's complex interplay of diverse weather systems. These settlers clearly had a good understanding of the normal weather systems for their respective areas, and how they affected their livelihoods.

Individual settlers were also interested in knowing how daily weather patterns varied from year to year on their properties, as well as between their properties and places nearby or farther afield. This was particularly evident in the Davidson diaries, which extended, albeit with breaks, from 1874 to 1900. With few exceptions, the handwriting suggested that almost all daily entries were written by Joseph Davidson, which ensured a consistent qualitative account of local weather systems and their impact on the farm. In his diary entries he frequently drew geographical and historical comparisons. The blocks along the bottom row of each part of Figure 2.4 (p. 54) represent geographical comparisons between the day's weather on Davidson's farm with, reading from left to right, that on neighbouring properties, across the region, and more widely. The blocks beside the left-hand axis of each part show time comparisons between what Davidson was experiencing with, reading from bottom to top, the weather at another time that season, during the previous twelve months and for more than one year. The remaining blocks in each part of the figure are for all remaining pairings of place and time. The words in each block of Figure 2.4(a) are direct quotations from Davidson's diaries to exemplify the depth of information available for this analysis. The 'three-prong' diagrams in Figure 2.4(b) depict numbers of diary entries relating to precipitation, temperature and wind, respectively. That figure demonstrates Davidson's personal interest in and awareness of geographical and historical patterns in the weather systems of southern New Zealand. As his local weather knowledge strengthened, he became more confident in describing a particular day as, for example, 'hotter than average', 'more like a spring than a winter day', or having 'the greatest flood in living memory'.

Davidson frequently recorded weather events that interfered with his regular farm activities or damaged buildings and fences. For the 36 years covered by his diaries I was able to distinguish between those entries that referred to (a) winds strong enough to damage buildings, fences and plants, or calm conditions when there was not even a breeze to dry sheep, grain and hay crops; (b) notably high or uncomfortably low temperatures; or (c) heavy rain with flooding or insufficient rain for crop and pasture growth. Strong northwest winds caused problems for Davidson and his family in spring and late summer, and southwest gales in winter exposed people and livestock to uncomfortably cold conditions. Davidson recorded calm spells that delayed the harvest when the crop was damp, as well as periods when runs of several days with intermittent showers had left the sheep too wet to shear. Although very hot weather made outside work difficult, persistently cold weather was a greater impediment. Worst of all were outbreaks of unseasonally cold southerly air, one of which he experienced on 28–30 January 1894 and another outbreak, which lasted several weeks, later that year. Heavy rain and flooding had several impacts on seasonal farming activities. Heavy rain from the easterly quarter in late summer and autumn usually resulted in flooded rivers and streams, and could isolate the farm for days at a time. The same was true after strong northwest winds and heavy rain in the back country had melted mountain snow packs and caused flood surges in the lowland reaches of the Waikaia River. From late spring to early autumn, too little rain could trigger the onset of drought, as in the six years starting 1888. During winter, when temperatures were too low for continuous pasture growth, too much or too little precipitation was seldom a major impediment to plant growth, but the late 1870s and the late 1890s – both of which were characterised by persistent rain, deep winter snow and unusually cold conditions – were challenging times for Davidson, as was the long run of unusually hot dry weather throughout 1880–82.

Weather events affected people, crops and livestock on the Davidson farm. Strong winds had an adverse effect on crops and pastures, a lesser impact on residents and their activities, and no reported effect on sheep and cattle (Figure 2.5, p. 56). Extremes of temperature were felt by residents and raised major concerns for the well-being of livestock when a combination of fog and cold in

	On the property	In the local area	In the region	Nationally and wider
More than 1 year	'We had one of the heaviest thunder storms about 9 o'clock p.m. there has been for years.' *(27 November 1877)*	'[The Mataura River] has not been as high since the big flood 9 years ago.' *(26 September 1878)*	'We have not seen so much fine weather at this season for some years.' *(24 April 1884)*	'There has been a heavy hot wave other side [of Tasman Sea], crops a failure both there and [in] New Zealand.' *(8 January 1898)*
Over past 9–12 months	'Weather keeping very dry – we started to cut grass seed with binder [but] found it too short to bind.' *(11 January 1886)*	'About the stormiest day we have had for some time.' *(24 April 1894)*	'Great complaints down-country about the crops being too short to cut.' [as a consequence of the prolonged drought] *(26 January 1886)*	No records
Current season	'Weather very mild for the season.' *(15 June 1880)*	'Had very bad weather, been very wet and cold for three weeks.' *(14 May 1885)*	'The weather is keeping fine for this time of year, in fact better than a couple of months ago.' *(3 May 1898)*	'By all accounts it has been very heavy rains from Dunedin northwards.' *(9 July 1885)*
That day		'Mailman could not cross the Mataura [River], too high.' *(4 July 1878)*	'Came on to rain here about 11 a.m. They had no rain at Riversdale.' *(17 April 1890)*	'There was a very heavy southerly buster this afternoon. It nearly knocked down my verandah. I think it has been felt all over the colony.' *(1 February 1898)*

FIGURE 2.4(a) Examples of Joseph Davidson's diary entries about the weather and its environmental and economic consequences. SOURCE: INFORMATION EXTRACTED FROM JOSEPH DAVIDSON'S DIARIES FOR HIS FARM AT SWITZERS / WAIKAIA IN NORTHERN SOUTHLAND; SEE APPENDIX NOTE 8 FOR DETAILS

winter or during lambing left tussocks and sown pastures coated with a layer of hoar frost. If they had sufficient supplementary feed for their sheep and cattle, then rural people were less concerned by winter snow than by several days of hard frost. Thermal stress on newborn lambs and damage to frost-tender crop plants such as potatoes, following outbreaks of very cold air in late spring or early summer, were also feared. Too little or too much rain in summer affected plant growth and harvest activities, and farm work was disrupted by heavy rain and flooding.

FIGURE 2.4(b) Joseph Davidson's geographical and historical comparisons of weather features.
SOURCE: INFORMATION EXTRACTED FROM JOSEPH DAVIDSON'S DIARIES FOR HIS FARM AT SWITZERS / WAIKAIA IN NORTHERN SOUTHLAND; SEE APPENDIX NOTE 8 FOR DETAILS

Across southern New Zealand, settlers soon recognised the effects of prevailing weather systems and recorded them in their diaries.[15] Availability of water was most important, followed by windiness and air temperature. An early lesson was that northwest weather is usually hot and dry in the lowlands but cold and wet in the back country. Another important lesson was that winter snow could be both a risk and a benefit to a flock of sheep. As the manager of Ida Valley Station wrote to the absentee landholder on 2 July 1891, 'June has been a trying month for them [i.e. the sheep]: first the snow, and since then terribly

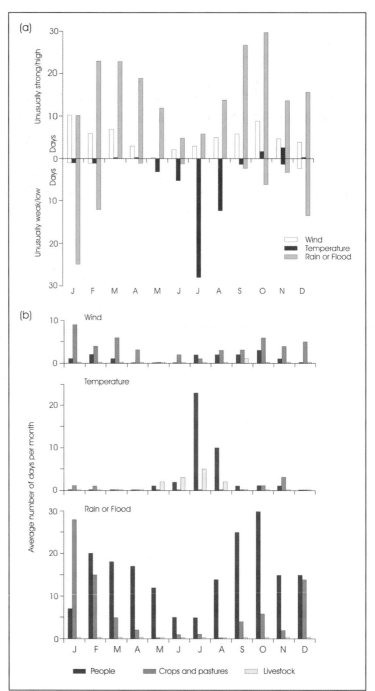

FIGURE 2.5 (a) Average number of days between 1874 and 1900 when: (i) unusually strong winds (above the horizontal line) or unusually still air (below that line), (ii) unusually high (above) or unusually low (below) air temperatures, and (iii) unusually heavy rain and flooding (above) or unusually little precipitation (below) gave rise to problems on the Davidson farm at Switzers / Waikaia in northern Southland. (b) On average, those months when wind, temperature and rain or flooding, respectively, caused significant problems for residents, crops, pastures or livestock. SOURCE: INFORMATION EXTRACTED FROM JOSEPH DAVIDSON'S DIARIES; SEE APPENDIX NOTE 8 FOR DETAILS.

hard frost and a great deal of fog. The fog I think must be worse for them than the snow, as every tussock is as hard as a board all day long, and never a blink of sun. There is only a dribble [of water] in the creeks, but of course this is all the better as it means that what little snow there was has gone into the ground. We want snow very badly for the [good of] the country.' In many respects, the weather system that followed a snowfall often caused most difficulty for the manager of Ida Valley Station: 'The heavy snow [in July] itself did very little harm (except to the [young] wethers) but the continuous wet cold miserable weather, with frequent snow storms, we had all August made the sheep lose condition more rapidly than I could have believed possible.' Clearly, a significant challenge to a settler was to learn to recognise and interpret weather signals, foretell what lay in store and manage his day-to-day operations accordingly.

INFORMAL WEATHER FORECASTS: FIRST STEPS

Amongst the clearest numerical signals of forthcoming weather are shifts in wind direction as well as trends in temperature and barometric pressure. Wind direction was occasionally reported, and most nineteenth-century farm and station diaries contained records of air temperature expressed in terms of human comfort but seldom in degrees Fahrenheit. On 23 March 1874, Joseph Davidson wrote that he had bought a barometer for £1 5s, and from then until the end of the century he infrequently recorded barometric pressure or noted if the 'glass' was stable, rising or falling. His diaries contained more than 300 daily observations of either barometric pressure or pressure trend, as well as descriptions of associated weather conditions. Initially, he recorded but did not interpret his readings, although several early entries showed Davidson's growing insights into the predictive value of daily barometric pressure readings. On 27 February 1876, for example, he wrote, 'Came on to rain in the evening. Wind from the Southward. Glass instead of going down kept rising to 29.80 [inches of mercury].' His interpretation of trends in barometric pressure also suggests an awareness of the weather guidelines set out in the front pages of commercially published diaries and the published advice of Robert FitzRoy.[16]

By the end of the century Davidson had grown adept at interpreting barometric pressure readings and trends, and on 18 April 1894 he recorded: 'The glass was very high at Set Fair, and it has rained since before noon. It got heavy towards night . . . . It looks as if we are going to have a flood.' The next day 'The Dome was at its highest about 8 a.m. [and] . . . . has not been so high since the Old Man Flood [of] 1870.' Three years later he was clearly basing his almost daily weather forecasts on barometer readings. Thus, on 19 August 1897 he wrote, 'Frost, but the glass is falling and there will likely be a change', and eight days later, 'The glass is up pretty high – looks as if we are going to have some settled weather.'

Two per cent of Davidson's diary references to barometric pressure related to 'very high' barometric pressure and associated weather systems: clear frosty days in winter, and either warm humid weather in late spring or rain and mild temperatures with winds from the easterly quarter in late summer. Nine per cent of the entries noted 'high' pressure, and the associated weather was fine frosty days in winter; clearing skies or strong northwest winds in spring; fine, dry, settled and occasionally hot weather in summer; and steady rain and stiff winds from the easterly quarter in late summer and early autumn. The latter weather systems were usually associated with flooding in local streams and rivers. Davidson reported a 'falling' trend in the 56 per cent of his diary entries that included barometric pressure, and he described 80 associated weather conditions: winds ranging from gentle breezes to southwesterly gales; precipitation ranging from heavy rain or snow in winter to mist or drizzle in summer; and temperatures varying between very cold conditions in winter, through mild days in spring and autumn, to occasionally very hot days in midsummer. Seven per cent of his entries related to 'low' barometric pressure, and were associated with heavy rain on the flats and snow showers in the hill country, squally weather in winter and spring, northwest gales in late spring and early summer, and either threatening or rainy weather in summer. Eight per cent of the entries concerned 'very low' pressures, and days with those reported readings were characterised by stormy weather with snow in winter, galeforce winds at any time of year, and heavy rain accompanied by flooding in summer and autumn. The remaining entries concerned 'rising' pressures and were associated with clear, often very frosty, weather and occasional falls of snow in winter and early spring, or fine warm weather in summer. Those findings are depicted in Figure 2.6 and show the overwhelming importance to Davidson of rapidly falling barometric pressure in any season as a clear warning of stormy weather and, therefore, the need to take special precautions when working

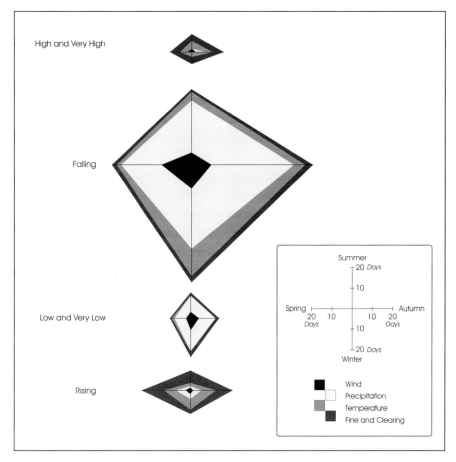

FIGURE 2.6 Features of weather systems associated with different barometric pressure trends, as reported by Joseph Davidson for his farm at Switzers / Waikaia between October 1874 and March 1899. SOURCE: INFORMATION COLLATED FROM JOSEPH DAVIDSON'S DIARIES; SEE APPENDIX NOTE 8 FOR DETAILS

with livestock and during harvesting. Rising barometric pressure, in contrast, told him that clearing skies and improving weather were on the way.

Davidson also recognised, interpreted and responded to visual weather signals: for example, 'Day fine – moon had a large circle around it – sign of wind in 48 hours' (26 September 1876); 'The weather pretty cold with a heavy coating of snow on the mountains – a sign of flood in spring' (28 July 1892); and 'A fine day, but glass falling – a large ring around the moon. Towards night the sky began to look like a coming storm' (3 April 1898). On 9 May 1898 he wrote in his diary, 'Has been raining a little last night. The day turned out dry with heavy clouds all round. The rock salt [he kept a large crystal of rock salt hanging by a length of thread in his kitchen to serve as a simple hygrometer] is dripping, a bad sign.'

FORECASTING THE WEATHER: SUCCESS BASED ON EXPERIENCE

With several years' experience on the property, supplemented by the records they kept of weather conditions and barometric pressure tendency, settlers grew more confident in their ability to forecast a change in the weather (Figure 2.7). Many forecast floods in large rivers after northwest gales in spring and early summer, and several perceived signals of adverse winter weather. Three of the farm and station diaries that covered at least five years contained detailed accounts of daily weather, and included what might be construed as simple weather forecasts. I extracted all entries where the writer had expressed an opinion about weather prospects during the day or days ahead, compared what he thought might happen with what eventuated, then calculated the proportion of correct forecasts for each year of the record.

All three settlers forecast the onset, continuance and cessation of rain or snow, wrote little about prospective air temperatures, but occasionally commented on the likely continuance of very hot or very cold conditions. Their individual rates of correctly forecasting weather systems one or two days ahead exceeded 80 per cent after about five years. Furthermore, their inferences about the consequences for their respective properties of episodes of strong wind and heavy rain in the headwaters of streams and rivers were mostly correct. With experience, Davidson learned how to forecast the normal range of weather conditions in northern Southland despite that area's complex weather patterns. In drier inland parts of north Otago, McMaster's task was simplified by the persistent interplay of northwest and southwest weather systems. The weather in lowland south Canterbury, where easterlies – often accompanied by rain – were common in summer, posed challenges to Gunn, but these were minor compared with the difficulties faced by Davidson.

Not all nineteenth- and early twentieth-century farmers and station holders appeared interested in forecasting the weather. For three decades, David Bryce at Lovells Flat in eastern Otago simply described each day's weather,

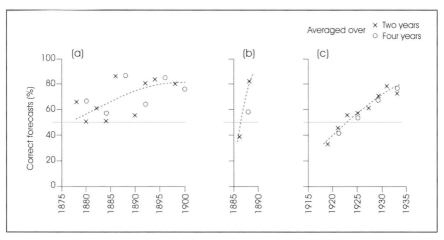

FIGURE 2.7 Three farmers' success in foretelling weather conditions one to three days ahead for their respective properties: the rates are averages for runs of two years and four years, respectively, over the period covered by each series of farm or station diaries: (a) Joseph Davidson's farm at Switzers / Waikaia, northern Southland; (b) the McMaster pastoral property at Tokarahi, inland north Otago; and (c) the Gunn family farm at Hook, south Canterbury. SOURCE: INFORMATION EXTRACTED FROM THE NAMED SETTLERS' DIARIES; SEE APPENDIX NOTE 8 FOR DETAILS

occasionally mentioned persistent drought conditions as well as major weather events and floods elsewhere in eastern Otago, and on just three occasions recorded his views about what lay ahead: 'It rained all day it is a very wet night [and] it looks like a flood' (6 August 1888); 'A very good day. It is very like rain' (17 June 1898); and 'A very good day; things are a good bit drier, but it is very like rain again' (15 July 1898).

An ability to forecast the next day's weather allowed settlers to speed up the grain harvest, move livestock to higher ground to protect them from rising water levels and prepare supplies of supplementary feed for their animals during winter. Davidson, Gunn and McMaster grew adept at forecasting weather conditions a day or two ahead, and were interested in knowing how closely their experience on a particular day related to what was happening elsewhere in the region or the country, as well as in similar conditions over the longer term. They acquired a sense of the magnitude and frequency of unusual weather events, but apparently not of cyclical changes in weather patterns over runs of several years.

We can discern evidence of the Southern Oscillation in the runs of weather observations recorded by settlers across southern New Zealand, but there is no reason to believe that farmers and station holders recognised these and other cyclical changes in weather and climate. Amongst their achievements, settlers learned to recognise variations in seasonal weather systems, and discovered how to forecast changes in the weather as well as the likely onset of flooding after snowmelt and heavy rain in the mountains. Furthermore, some perceived the signals of adverse weather and managed their operations on the land accordingly.

PROSPECTS FOR FORMAL WEATHER FORECASTS

In its 14 December 1860 edition, the *Southern Cross* printed text drawn from *A Manual of the Barometer* by Robert FitzRoy, which had earlier been published by the Board of Trade in London. Under the headline, 'How to Foretell the Weather', it told interested readers how to interpret observations of barometric pressure, air temperature, humidity, cloud cover and type, wind strength and direction. On 17 May 1862 the *Otago Witness* reprinted a piece from the British press about the benefits that would flow from a weather forecasting system able to provide two or three days' warning of an impending storm. At that time, a committee of the Royal Society in London had concluded that 'weather prediction . . . was not the business of a "strictly scientific body"',[17] suggesting it was an art rather than a science, but nevertheless the benefits of observation and the use of informal and semi-scientific notation as well as good-quality measuring instruments were coming to the fore. The primary need for a network of meteorological observatories was picked up in southern New Zealand on 3 June 1862, when the *Otago Daily Times* published an approving notice about the recent establishment by the British Board of Trade of a meteorological office: 'We cannot too urgently direct [this to] the attention of the Central [New Zealand] Legislature.' The benefits, it argued, would be felt by mariners as well as by people on the land. The newspaper proposed that a formal request should be made to Robert FitzRoy, director of the Meteorological Office and a former governor of New Zealand, to send a competent person to New Zealand to advise on issues such as measurement and instrumentation.[18] The article concluded that a major use for the newly installed telegraphic system would be for transmitting meteorological information between population centres.

Other local newspapers joined the hunt, and *The Press* on 6 January 1864 printed a letter written by FitzRoy and first published in *The Times* of London in which he laid out the principles of forecasting stormy weather. Three years later, on 15 March 1867, the *Otago Daily Times* reprinted a piece from the *London Review* under the banner 'Weather Wisdom' that included this strikingly prescient opinion: 'that weather may be forecast is certain thanks to the knowledge of scientific meteorology and the electric telegraph'. The writer further suggested that forecasts could soon be made for one or two days ahead, at least, and reported the hypotheses of John Herschell and Alexander von Humboldt that the moon was the primary influence on the earth's weather systems. The following year, on 28 February 1868, in announcing the recent publication of a two-volume compilation of meteorological statistics for Italy, the editor of *The Press* made this guarded endorsement: 'the fact that that country is situated in the same latitude in the northern hemisphere that New Zealand is in the southern renders them [the published statistics] exceedingly interesting'.

On 16 January 1864 the *Lyttelton Times* reprinted a short piece from the *Hobart Town Mercury*, which had been taken from the *Nautical Magazine*, under the headline 'Saxby's Weather System'. It gave dates between July 1863 and March 1864 'on which the weather may reasonably be suspected as liable to change, most probably towards high winds or lower temperature, being especially periods of atmospheric disturbance'. Stephen Saxby (1804–83) had proposed a system for predicting 'dangerous periods' for storms, based on the moon's position relative to the equator, and in 1862 he published details in *Foretelling Weather*. A second edition, with the title *Saxby's Weather System*, was published in 1864. His system was well known in southern New Zealand and frequently cited in newspaper articles and government reports. It did not, however, enjoy FitzRoy's approval.

Nevertheless, in New Zealand such theories were accepted for consideration. In its edition for 29 March 1867, *The Press* reprinted sections of an article that had originally been published by the *London Review*, to which it added favourable comments about the Herschel and von Humboldt theories of lunar influence on the weather,[19] as well as the capacity of some animals to give warning of storms. The *Otago Daily Times* in its 11 August 1868 edition printed the inaugural address delivered to members of the New Zealand Institute by Sir George Bowen, a speech notable for its general endorsement of science and the speaker's high expectations of scientific meteorology: 'It might almost be said that every colonist in a new and unexplored country is, unconsciously, more or less a scientific observer . . . . The study of meteorology will prove of much practical benefit in these tempestuous latitudes; for the discoveries of Sir W. Reed and his followers have enabled science to encircle with definite laws the apparently capricious phenomena of the atmosphere.' In October 1869, *The Press* reprinted a paragraph from the *Southern Cross* about an unusually high tide in the Hokianga Harbour during one of Saxby's predicted 'dangerous' days, including these words: 'It appears that Saxby's predictions, which have caused so much uneasiness in the minds of many throughout New Zealand, have been fulfilled in many places.'

While debates about science and the efficacy of meteorology went back and forth, rural people had to make do as best they could.

PUTTING LOCAL WEATHER KNOWLEDGE TO GOOD USE

Local weather conditions had a marked impact on farm planning and practice, as an entry in the letter book for Ida Valley Station in July 1879 indicates: 'The sheep will no doubt have suffered all over [the district] from the severe winter we have had, but not so much from the snow as from the severe frost which followed a partial thaw of the first snow we had and which, on the flat at least, converted the snow into ice . . . . From the arguments in the [news]papers and the appearance of the Old Man, Dunstan and Hawkdun Ranges, we have been let off lightly so far . . . . It strikes me that this annual impossibility of ploughing – in the winter for the frost, and in summer and autumn for the drought – should be a bar on farming on any large scale in this district.'[20]

William Shirres, who managed Aviemore Station, was of much the same opinion when he dismissed a neighbouring runholder's 'extravagant theory' to sow 400 acres in turnips so that he could carry 9000 sheep on a property where 5000 struck Shirres as closer to the mark. He had advised his neighbour 'to consult someone who knew the country well' before embarking on this improvement.[21] This interchange finds its echo in the words of two Lincolnshire farmers, who had recently visited New Zealand to assess opportunities for intending rural settlers, when they wrote about the recently founded Lincoln College: 'The object of the institution is to enable those who take advantage of

it to acquire a thorough knowledge of practical and scientific agriculture . . . .
[for] it is confessed by many of the farmers that a more scientific knowledge of
farming is greatly needed among them.'[22]

After a few years on the land, settlers began to understand how environmental conditions could change over short distances and under different
plant covers. Thus, on 1 October 1886, towards the end of a long drought,
William Shirres wrote to a director of the finance company in London that held
the mortgage on Aviemore Station: 'We have had a most favourable winter.
McAughter (shepherd) says he has never seen the ewes look so well on this
country before – the [live]stock along the coast, however, on English grass paddocks have suffered most severely. As you will remember, the paddocks were
terribly bare when you were here, so that there was bound to be a scarcity of
feed when winter came on. To add to the misfortunes of farmers, there was a fall
of nearly six inches of snow all along the coast just at lambing time.'[23]

With their diaries and letters to guide us, we can track what and how early
settlers learned about the environments of southern New Zealand, particularly
the beginning and end of the growing season; the time when it was appropriate
to sow frost-tender plants; soil conditions best suited to root crops and vegetables; and varieties of pasture plants, root crops and grains most likely to
do well on their properties. With respect to the latter, Joseph Davidson wrote
in his diary on 9 March 1883 that he had 'Put Canadian oats in chaff stack as I
find they shake out very easy; very good oats but no use in this district as they
cannot stand the wind and weather like Short Tartarian and Sparrowbill.' He
also found that a machine imported from the United States for sowing grain
was ineffective in the windy conditions of northern Southland.

In time, settlers learned about the geographically and historically variable
weather systems and climates of their properties and the long-term implications of these conditions for animal welfare and economic production. As
the manager of Ida Valley Station wrote to the non-resident landholder on 19
August 1875:

I have previously over-estimated the carrying capacity of this Run, in my opinion, and now I think otherwise I feel it my duty to mention it at once. I may
mention that I have lately ridden over a considerable portion of the Galloway
Run, where there is absolutely no feed whatever on thousands of acres, and as
far as I can judge little chance of there ever being any permanent improvement

in the country without total rest [from grazing] for a season at least. The gradual drying of all the country hereabouts that has a Nor-west exposure has I think a great deal to do in recovering country [by regrowth of palatable and other plants] after being overstocked.

The depth of the manager's knowledge of local environmental conditions and their likely effects on the land and livestock are manifest in his letter book. In spring 1879, on 15 September, he wrote, 'I also wish to shear earlier [than 15 December] to get the scouring finished before the water failed, as it always does about Xmas. You know how much cleaner and more free from dust the wool is when shorn moderately early, and what a difference even a week's earlier shearing makes for their condition for winter in this climate.' And on 18 July 1888 he concluded that 'though an open winter like the present may be considered a good one for stock, I prefer the hard dry winters that are usual here. There is this in favour of a good spring this year. The ground has got a soaking for the first time almost for 2 years as the snow has been allowed to soak into the ground, which was not frost bound as usual.' Years later, on 9 December 1892, he wrote, 'It is the old story, wishing for rain. I was over at Puketoe on Friday (2nd) and there was very heavy rain there, but none this side of the hills.'

During the second half of the nineteenth century, weather conditions remained a topic of prime interest to pioneer families until their experience and the advent of simple weather forecasts enabled them to judge when to move livestock to safer sites, reschedule harvesting and shearing tasks, and take other appropriate actions to reduce the likely impact of an unusual weather event. Experience on the land also allowed them to make far-reaching decisions about stocking densities; which tree, shrub and herb species to plant; and how best to use the different topographic features of their properties. In their letters to absentee landholders, managers of large farms and stations were reporting such matters as early as the 1870s, and there is no reason to believe that the residents of other properties were less sensitive to the dynamic environments of southern New Zealand.

CHAPTER THREE

Exceptional Challenges
Flood and Drought, Ice and Snow

Settlers faced a difficult task learning how to identify reliable signals for the normal range of weather conditions, interpret them and take appropriate action, but the task of reliably forecasting the onset of exceptional conditions – widespread flooding, prolonged drought, a week or longer of sub-zero weather, and snow in deep drifts – lay beyond their ability or, indeed, that of anyone else in New Zealand during the second half of the nineteenth century. However, because such events could recur several times during an individual's lifetime and significantly affect the economy of a farm or station, it was essential for settlers to be prepared for spells of exceptional weather.

A pervasive sense of optimism led settlers in southern New Zealand to transform their properties from a 'howling wilderness' to a humanised rural landscape. They used a kitset of implements, tools, plants, lumber, fence posts and galvanised wire with which they cultivated hedges and erected post-and-wire fences to regulate where livestock grazed; ploughed the soil and sowed pasture and crop plants; and established plantations of tall trees to mark

Snow at St Bathans, Central Otago, *circa* 1908.

property and field boundaries, provide fuel and lumber, and protect homestead, livestock, crops and pastures from strong winds and winter cold. For
some individuals the experience of ordinary weather made them think exceptional weather was never going to occur:

> There is plenty of snow among the hills, but it falls very seldom on the
> [Canterbury] plains. And when it does it disappears in a few hours. It often
> freezes at night, but never severely enough to remain in the ground through the
> day, and in frosty weather the days are invariably bright and clear .... [Strong
> northwest weather] is frequently succeeded by a sudden shift to the south-west,
> accompanied sometimes by a heavy rain for a few hours. One consequence of
> this, and it is an important one, is that the country is not liable to droughts. It is

not dry long enough together to make the grass burn; and there is generally rain enough through the summer, to keep it growing.[1]

Optimistic statements like that were common in the handbooks directed to intending migrants, and the author of one from 1880 ascribed 'the healthful character of the New Zealand climate' to 'the clear elastic atmosphere', the balance between precipitation and evaporation, the absence of extreme temperatures, an abundance of running water and the lack of pestilential swamps.[2] In the same year, a commissioned report on the agricultural conditions and prospects of New Zealand was published.[3] Its authors, S. Grant and J. S. Foster, who referred to themselves as 'delegates to the Colony from the tenant farmers of Lincolnshire', gave a generally informative account of farming and rural environments in the later years of the nineteenth century. They made sensible suggestions about stocking densities for farms on the dry soils of the Canterbury Plains, and warned intending settlers to inquire whether a property was likely to experience flooding before arranging to buy it; yet they did not mention the widespread flooding and major snowstorms of the late 1860s. In his handbook, first published in 1879 and revised for re-publication the following year, the eminent New Zealand scientist James Hector included extensive tables of meteorological statistics from a small number of well-equipped weather stations distributed across the country.[4] He described normal seasonal weather, but referred only once to drought and not at all to floods or snow, despite the appalling weather of 1867 and 1868 that had affected much of eastern and southern New Zealand.

Situated almost as far as it was possible to be from the British markets for their wool and hides, settlers in southern New Zealand had to produce quality outputs at the lowest possible price. This required individual landholders to run large flocks of sheep, which was only possible on highly productive, nutritious pastures. The high stocking densities required for economic viability added to settlers' vulnerability to extreme weather events, and individual landholders soon learned that they needed to identify signals of adverse weather systems and to set in place procedures to mitigate their effects. To those ends, settlers planted trees and shrubs to shelter livestock from strong winds, and erected low dams across creeks and small streams to accumulate supplies of water for livestock and irrigation during periods of summer drought, as well as to meet the demands of wool scouring after shearing in summer. A landholder usually had sufficient water on hand to cover the normal range of seasonal variations in supply and demand, but the prospect of exceptional weather required a longer-term perspective. One strategy, although not widely followed at the time, was to stock the entire property with sufficient animals for them to survive bad, but not the worst possible, environmental conditions somewhere on the farm or station. A more common strategy was to treat the property as a suite of distinctive, small-area environments, place fences around them, and stock each with just sufficient animals for them to survive environmental conditions outside the normal range. The most common strategy was to keep as many animals as the landholder could afford, attempt to lay up sufficient supplies of high-energy feed to tide them over summer drought and winter cold, and hope that financial losses incurred during bad years would be covered by profits during the good years.

It would be difficult to overestimate the impact of exceptional environmental conditions on a pioneer farm or station, especially in those parts of the country where major weather events could be experienced every three or four years. Strong winds, widespread flooding and heavy snowfalls during the second half of the 1860s were times of trial for pioneer settlers in southern New Zealand. Only when the country had functioning road, rail and maritime transport links and a reliable telegraph system did farm and station holders receive timely warnings about the magnitude and geographical extent of adverse environment conditions. Several settlers' diaries reported recent rainfall on neighbouring properties, but none on theirs. Others compared presence and depth of snow on their own and their neighbours' properties. While it is unclear whether these were reflections of the landholder's interest in the weather, a warning of things to come or evidence of shared suffering, comparisons such as these provided geographical context and were important when the opportunity arose to buy all or part of a nearby property. Station holders were also kept informed by local and national publications about the consequences of exceptional weather across the Tasman Sea and farther afield. The scientific discipline of meteorology was nascent during the first two decades of organised settlement in southern New Zealand, and relations between moon and earth were widely believed to trigger episodes of stormy weather. Published long-range forecasts of weather and climatic conditions were available, and the interest of rural people in knowing the likelihood of an extreme weather event was evident in the letters columns of local and regional newspapers.

FLOOD

In its 17 May 1862 edition, the *Otago Daily Times* reported that a violent storm had swept down the eastern flank of the North Island from Auckland to Cook Strait, thence south into Marlborough, Canterbury and north Otago. It was one of several such storms to affect southern New Zealand that decade. By then, settlers were aware of the links between wind direction and seasonal snow-melt in the mountains, and were beginning to forecast the consequences for their properties and the district. An example comes from the 20 August 1862 edition of the *Otago Daily Times*, in which its Southland correspondent wrote, 'The snow [fed] rivers will rise shortly from the melting of the large quanti-ties of snow which have fallen this winter on the higher hills at their sources. Great caution should be exercised by strangers when crossing them, either on foot or with drays . . . . People coming from Otago should be careful not to cross unless assured that it would be safe to do so by someone thoroughly acquainted not only with the fords but also the state of the river.' Later that year, in its 4 October edition, the *Otago Witness* quoted from a private letter written on 24 September by a settler in the Manuherikia district: 'The wind is still blowing very hard. Very probably tomorrow or the next day the Norwester will termi-nate by veering round to the Sou-west, and we shall have cold winds and snow on the mountains. As soon as the change of wind takes place, the waters will abate, and in the course of four or five days many of them [i.e. the goldminers] will get to work again.' Warnings of adverse weather and floods were facilitated by tele-graphic connections between population centres, and on 19 November 1866 *The Press* reported the likelihood of a sharp rise in the level of the Waimakariri River, thanks to information telegraphed to Christchurch from Bealey, a small settlement deep in the Southern Alps, which gave between four and 24 hours' warning of flooding along the stretch of the river where it crossed the plains.

On 8 January 1868, *The Press* reported that the Rakaia River, swollen by the recent rains, was flowing bank to bank and eroding its northern flanks. Water levels were described as higher than those of the flood at Christmas 1865. January was wet across much of New Zealand, and on the seventeenth *The Press* printed a report from its correspondent in the Otago goldfields about the impact of several months of heavy rainfall and flooding on mining operations. So serious and widespread was flooding during the first six weeks of 1868 that *The Press* published a series of long and detailed articles for the information of its readers. On 5 February it reported a fresh in the Avon River and widespread flooding throughout Christchurch. The main road bridge over the Selwyn River had been washed away and low-lying land on the flanks of Lake Ellesmere was inundated. The next issue described the onset of warm southeasterly weather over Banks Peninsula during the weekend, followed by persistently heavy rain and flooding in rivers and streams, submerging low-lying land and isolating farmhouses. Even Māori settlements known to have escaped flood damage since the early 1850s were affected. Elsewhere, the inland forestry town of Oxford was isolated; the Ashley River broke its banks; flood waters 'came down like a huge wave' in the Hurunui River; Blenheim was 'one sheet of water' two feet deep; and Wellington was experiencing wet stormy weather. News of adverse weather throughout the eastern South Island was published in the 8 February 1868 edition of the *West Coast Times*, which reported, 'From Christchurch we learn that very heavy floods have prevailed throughout the whole district [Canterbury], the weather being of a quite exceptional character. Several of the bridges across the rivers have been washed away.'

The adverse impact of flooding was not confined to the eastern lowlands. In its 10 February edition, the *West Coast Times* reported 'floods and landslips at Porter's Pass', and observed that 'it is difficult to tell where has been the greatest amount of suffering. The farmers are ruined, [pastoral] station masters are very heavy losers, and tradesmen of all kinds have to count their losses . . . . communications being cut off by bridges, wires and road, so we only know of the destruction around us . . . . boats are plying to and from Colombo Street to the Post Office [in Christchurch].' Four days later the same newspaper reported 34 known deaths in the recent floods. Despite the strong language of this report, it soon became evident that major damage was confined to the Canterbury Plains and eastern foothills.

The editor of *The Press*, echoing the opinions of his peers, questioned the wisdom of siting human settlements on the banks of large rivers, especially in areas that should be reserved for ponding in the event of a large flood, and on 10 February 1868 that newspaper published extracts from the log of the *Canterbury*, which had left Wellington on the eve of the storm and travelled through it to Lyttelton. This report must have given readers a sense of the magnitude and extent of the event, and indicated the weather system's sub-tropical origins. *The Press* confirmed that while the storm and the attendant rain had affected the eastern lowlands, the weather system had not penetrated far into the mountains or the Mackenzie basin. The editor wrote, 'Nothing like this [storm] had ever been seen before in this country, in the twenty years during which it has been occupied, and it may be another fifty years before such a phenomenon may occur again.' He called for the province's rivers to be 'restrained' and willows planted alongside them to protect their banks from erosion. By 10 February a clearer picture was emerging of the recent floods and their effects on the land and people of Canterbury. Temuka was inundated when the Waihi, Temuka and Opihi rivers – all of them rising in the eastern foothills, and primarily rain fed – rose in a 'terrific flood'. Road bridges between Timaru and the Waitaki River had been washed away, settlements were isolated and rural property was damaged, inspiring a perceptive correspondent to declare in a letter to the editor in the same edition: 'My motto is, assist nature to help herself in all cases, [and] not obstruct her.'

Despite major damage to infrastructure and sorely disrupted communications by land, news of the flood filtered between Christchurch and places farther south. On 4 February the *Otago Daily Times* reported 'the extreme inclemency of the weather' in Dunedin, which started with an 'immense downpour of rain the previous day'. A report drawn from the 5 February edition of the *Otago Daily Times* referred to heavy rain 'causing severe floods throughout the whole of the province, apparently', and mentioned flooding in the Leith Stream that ran through Dunedin, landslides along the coach road to the north, flood damage to the botanical gardens, and rising water in the Taieri, Tokomairiro and Molyneux rivers. When invited to search their memories, readers informed the editor of *The Press* on 12 February that the flood could not 'be accounted for by the experience of the oldest settlers', although one correspondent proposed the unlikely explanation of recent earthquakes disturbing large alpine snow packs that remained from the previous winter. The lessons and financial cost of the February floods were quickly recognised, and one correspondent who farmed near Lake Ellesmere wrote to *The Press* on 27 February about the loss of 261 acres of wheat, 106 acres of oats and 28 acres of barley, as well as £250–£300 damage to fences on his farm.

From then until the middle of the month, the *Otago Daily Times* published several long pieces about the floods in Otago and reprinted blocks of text from *The Press* about the situation in Canterbury, including advice to farmers and the authorities responsible to plant willows along the banks of flood-prone rivers and streams. On 5 March the *Otago Daily Times* described the recent floods as serious in Otago and Southland, and sagely concluded, 'The first impression was that the losses of the settlers were enormous, and almost irreplaceable, but subsequent accounts show that they were very much over-estimated.' Four days later the *Otago Witness* reported the impact of the recent heavy rains and severe flooding in Dunedin, along the main roads north and south of the city, and in the goldfield towns of Lawrence and Clyde. It described the floods as without comparison to those experienced by settlers over the past two decades. They had inundated almost the full length of the Taieri Plain, caused breaks in telegraphic connections to the north and south, and were widespread throughout eastern Otago. The editor called for replacement bridges to be erected as high as possible and without piles in the channel wherever feasible, for it had been found during the recent floods that low structures on closely spaced piles trapped debris and caused flood water to back up and damage them. The following week, on 15 February, it reprinted an editorial from the *Otago Daily Times* calling for buried telegraph lines as a way to guard against further breaks in communications during a severe storm and, hence, allow warning of any such event.

Widespread flooding during the late summer of 1868 was the foretaste of a year of stormy weather. On 3 June, according to the following day's edition of *The Press*, a southwesterly storm resulted in 'a heavier fall of snow than had occurred here [Christchurch] for years . . . . and the town and adjacent country[side] were soon covered with a white robe'. The weeks that followed saw hard frosts throughout the province, heavy rain in the hill country and flooding in rivers that rose in the eastern foothills of the Alps. Commentators again expressed concern about the ongoing drainage of swamps for farmland leaving rivers to carry more water than their channels could safely accommodate, but the greatest worries were the widespread incidence of flooding – albeit less severe than in February – and the perceived threat to Christchurch posed by the Waimakariri River overflowing its banks or changing its course to flow down the shallow depression normally followed by the Avon River. In many respects, the floods of June 1868 were more serious than those of four months earlier, for they affected virtually the entire eastern flanks of the South Island as well as the inter-montane basins. On 15 June the *Otago Daily Times* published a long article under the headline 'Heavy Gale and Floods': 'The heavy rain of Friday [12 June] has resulted in a flood which, in some parts of the Province [of Otago] at least, has been more severe than that which occurred four or five months ago

. . . . That flood was fortunately confined to a relatively narrow strip of country from the seaboard, but the present one, it is to be feared, has extended over nearly the whole of the Province. The Clutha [River], which was scarcely at all affected on the former occasion, has now been greatly flooded.' The newspaper also reported major floods in the Taieri Plain and the Tokomairiro lowlands, damage to coastal structures and landforms, and shipwrecks.

In its 16 June 1868 edition, the *Southland News* reported 'a fearful storm in Timaru on Saturday' accompanied by 24 hours of continuous rainfall, flooding in already high rivers and streams, and mountainous seas. It also described southerly and southeasterly storms in north Otago, floods at Balclutha and continuous rain in Lawrence. Two days later it observed that February 1868 had been a month 'in which a series of extraordinary storms prevailed on the East Coast, causing fearful floods and great deprivation. From this visitation, Southland was happily exempt.' That province did not, however, escape so lightly in June, and on 26 June the *Otago Daily Times* quoted from a report written by the Switzers (present-day Waikaia) correspondent of the *Lake Wakatip Mail* and published by it on 9 June: 'The winter has set in with a vengeance. We have had a snowstorm that has continued four days, blocking up the roads, stopping traffic, and far worse than this, seriously impeding mining operations. I now begin to believe the Maori prophesy of August 1867 – that New Zealand will be visited by a succession of storms and rain alternately and continuously for one year – will be fulfilled.' In its 1 November 1868 edition, *The Press* reported major flooding on 17 and 18 October in the Australian state of Victoria after a long period of drought. It showed New Zealand readers that they were not alone in having experienced extreme weather that year.

In northern Southland on 10 November 1891 Joseph Davidson recorded in his diary: 'About 5 a.m. the Dome Creek was bank-to-bank, washing the fence down and changing its channel in many places. The lower parts of the paddocks and my potatoes are under water, the [Waikaia] River was very high towards evening . . . . [causing] a great slaughter among the rabbits.' Four years later, in September 1895, Archibald Morton, then the manager of Haldon Station in the Mackenzie Country, wrote to entrepreneur James Preston, who held the lease, 'We have had a dreadful time with [flood] water about the house. There is a heavy stream flowing down still between the stable and the house. The fences are all washed away in every place there is water. At the dam a gap is washed clean out and spouts where the water comes in are washed right out.' In late

December he wrote, 'We have had a dreadful time of rain. It has been raining every day or two since I got back from the [Timaru Agricultural and Pastoral] Show . . . . the weather is wet, and they have had bad weather down country.' Conditions were so bad that shearing was delayed throughout much of south Canterbury, north Otago and the Waitaki Valley, and Morton wrote to Preston, 'I finished shearing last month [December]. A lot of my shearers never turned up. The weather was so bad down country that they could not get cut out in time . . . . Shearers are not to be had about Timaru.'[5]

There were further episodes of flooding during the remaining years of the nineteenth century, and on 27 September 1898 Archibald Morton wrote to the absentee landholder, James Preston, 'I had a job to save the dam. It [the flood waters] swept away the scrub work . . . . There has not been so much water about since the flood of '95.'[6]

Throughout the second half of the nineteenth century, floods remained a major worry for settlers until well-engineered bridges could be built and reliable telegraphic communications installed. The plea written on 2 May 1891 by J. Rattray of Dunedin to his valued client, James Preston, after agreeing to a further injection of finance to enable the purchase of Ben Ohau Station, encapsulates the concern then felt by many people: 'I have paid the £1000 and wish to have the satisfaction and pleasure of seeing you through all your engagements – which I do not doubt two or three years will accomplish in your customary solid and honourable manner. One thing is very necessary. You must run no risks with yourself in night journeys or crossing rivers or other rash ways.'[7]

Floods continued to attract great interest amongst people in rural areas at the turn of the century, and were reported widely. On 7 March 1904, for example, the manager of the Gore branch of the New Zealand Loan & Mercantile Agency wrote to the regional manager in Invercargill:

After an exceptionally fine dry season which allowed for most of the grass in the district being cut and stacked, the weather broke on Saturday evening, February 27th and the rivers and streams rose slightly but fell again. The ground became thoroughly saturated, and when a heavy rain set in again on Tuesday night the Mataura and Waikaka rivers, the Waimea, Otamita and other streams rose rapidly and on Friday were in heavy flood. The rain cleared off on Friday morning, but since that time the weather has been hot and muggy, with no drying winds to dry the grain . . . . The worst damage was done on river flats, and a large quantity of oats in sheaves was carried away from fields on the river banks. On the flat lands, where the flood water lay, a considerable amount of damage was done to stacked oats and in some cases stacks were broken down and portions washed away. Further damage will be done with silt ... where fields of good pasture will be spoiled, being covered with silt to such an extent that this season's feeding will be lost, it being too late to expect much further growth.... A considerable amount of damage has been done to fences. The worst damage has been done on the banks of the Mataura River, the Waimea and Otamita streams, and in the Waikaka valley.[8]

DROUGHT, DUST AND HEAT WAVE

Far from finding the equable, temperate climate described in handbooks published for their information, newcomers discovered that southern New Zealand is subject to occasional episodes of extreme weather, including protracted periods of sub-optimal rainfall, even drought. In its 19 December 1862 edition, the *Otago Witness* published the following report from its correspondent in Canterbury: 'The weather continues terribly warm and dry. The farmers are calling loudly for rain, the want of which is apparent on every description of crop, especially the hay crop, which this year is one of the lightest known here. Such a thing, too, as fat beef or mutton is unobtainable on account of the scantiness of the pasture. Christchurch in such weather is a most disagreeable place of residence.... for the dust is flying.' That issue of the newspaper also reported severe drought in Queensland and bush fires in Otago. Later in the decade, on 2 May 1868, the *Otago Daily Times* published the following brief item taken from the *Bruce Herald*: 'The drought in the Tuapeka District still continues to the serious inconvenience of every inhabitant. Not only is the mining industry partially suspended but domestic operations and healthy ablutions are crippled for want of water.' Trying environmental conditions were also experienced farther south, as the 17 November 1868 issue of the *Southland News* reported: 'The dust lay thick in the streets [of Invercargill], but the wind rarely blew hard enough to raise it, although once or twice there was a regular "brickfielder" [dust storm]. This situation, with persistent winds from the northerly quadrant, reached its climax on Friday evening [12 November] when heavy rain fell.' On 19 January 1869, *The Press* reported that the 1868 summer had been a time of drought in Europe while India had been deluged by monsoon rains, but a rational explanation for this would not be possible for more than a century. Drought conditions persisted throughout the summer of 1868–69, and on 25 January 1869, *The Press* printed a short piece from the *Bruce Herald* about the miserable appearance of oat crops in the Tokomairiro district of east Otago. It followed that with an item on 16 February, also taken from the *Bruce Herald*: 'The general appearance of the crops throughout the [Waikouaiti] district does not promise the usual average yields, the long drought which prevailed when rain was most needed having greatly retarded their growth. On the average, however, we believe the crop will compare favourably with other districts.'

The decade of the 1860s was marked by heavy rain and flooding, but the 1880s and early 1890s were times of widespread and persistent drought in southern New Zealand. On 5 January 1882, Joseph Davidson wrote in his diary, 'Weather keeping very dry – too dry for the turnips; [I am] afraid they will be a very poor crop on new ground.' One month later he wrote, 'Crops very backward this season. The rain [five days ago] has put the turnips forward.' On 14 March 1886 William Shirres, manager of Avicmore Station in the upper Waitaki Valley, wrote to the financial backers in London that 'Things in New Zealand are in a bad way. This has been the driest summer known by Europeans. Until lately, all the pastures down country have been burnt up, and the [live]stock have suffered severely.' He reported upland pastures had put on more growth, but that the situation had prompted Otago runholders to petition government for a reduction in rent as a drought relief measure.[9]

In a letter to his father, written on 25 January 1887, James Preston described the worsening situation at Longlands Station in the Maniototo: 'The weather is very dry. A great quantity of the wool is sandy [from the blown dust], and this will be quite hard to scour. The dry sheep are in good condition, and the lambs are pretty good, but the want of water is telling on them now.' Two years later, on 6 August 1889, the manager of Ida Valley Station wrote to the absentee landholder, 'Heavy rain (snow on the hills) fell last Friday, the first I am afraid to say for how long. Before it, the water in the wells was lower than it is in the driest summer. This is surely good proof of the drought I have written so much about lately. We shall be more than ordinarily anxious about the weather for the next month. If good, I believe we should come out of the terrible autumn and winter better than expected.... Some of the young sheep are looking very bad, but they have been running all winter on the low ground where there was really not a bite [of grass].' The situation at Ida Valley Station remained serious throughout spring, and on 1 November 1889 the manager wrote, 'The ewes are still in poor condition and the country dreadfully bare. The young sheep have improved considerably, and if rain would only come I think they would clip [well] .... Still no rain – I really don't know what this country is coming to .... This continual drought is nearly past bearing .... this is practically the third year [of drought] here. It is a wonder things are even no worse than they are.'

On several occasions during March 1891, Preston recorded in his diaries that employees at Longlands Station were irrigating cropland because the soil was too dry for sown pastures and the young turnips. Two years later, on 1 March 1893, Archibald Morton, who managed Haldon Station for Preston, wrote to the latter 'I am horribly sick of writing year after year about this dry weather as you must be tired of reading about it, but we [are] nearly as bad off this year as ever. The high country is (baring scarcity of water) all right but the lower parts are completely parched and any grass on them [is] as wizened and brittle as straw.'[10] Two weeks later he reported, 'there has been no pasture growth for two months .... I had hoped to get in some rye [probably ryegrass seed] for winter feed before now, but a plough won't enter the ground, and is now almost too late.' In May 1895 he wrote to Preston, 'The state of the lower country here is deplorable, and what sheep are to do for the next two or three months I don't know. All the country about here lying to the north is in the same condition, and in some places sheep are dying already.' In an effort to reduce livestock numbers he had tried to sell 500 ewes and 200 wethers at a reserve price of 3s 6d each, but did not receive bids for any of them. The considerable quantity of hay on hand, however, eased his looming sense of anxiety, although two years later, on 8 May 1897, conditions had still not greatly improved: 'We are having splendid weather, never saw better at this time of year since I came to Haldon .... [but] The ground is dreadfully dry. I never saw the back [stock] yards [as] cut up as they did this year.' At the end of the year he wrote again to Preston, 'I am afraid unless rain comes there won't be much feed for the run in winter. I suppose there is no hope of Mr Pringle giving up [the adjoining] ewe run. It is very awkward without it.'

The Mackenzie Country was not the only part of southern New Zealand then experiencing prolonged drought. According to Joseph Davidson, writing in his diary on 8 January 1898, even normally damp parts of the country like northern Southland were affected: 'A fine day, but glass falling a little ....

I see there have been heavy bush fires on New Year's Day: Round Hill, Colenso and Nelson, also very heavy fires in Tasmania. There has been a heavy hot wave other side [of the Tasman Sea], crops a great failure there and NZ. They are reaping in Canterbury – crops short [in stature?].'

ICE AND SNOW

In its 1 November 1865 edition, the *Otago Witness* published a short report from its correspondent in the Tokomairiro district of south Otago: 'Today in snow arrayed stern winter rules the ravaged plain. And the memory of that unimpeachable authority, "the oldest resident" fails to recall an instance of such extreme severity of weather as this season of the year. As far as my own experience goes, I can only remember one instance at all parallel, which was some seven or eight years ago, when we had a fall of snow on the first of November. On Monday it began to snow, and there have been occasional showers of it ever since, just sufficient to keep the ground white.' News of the widespread snowfalls between late July and early August 1867 even reached family members living in England, and Abraham Taylor of Dover Mill wrote to the Preston family on 31 October, 'We are very sorry to hear of the loss you are likely to have from the snow. We had no idea that snow had ever fallen so thick with you, except on the very high mountains. It would appear that the Leicester breed [of sheep] is too fine bred and too large for the Run and for the native grasses, and requires far more care. But I suppose that the cross breed will yield far more wool, and [be] more proper for combing purposes.'

There was more of the same the following year, and on 9 July 1868 the *Otago Daily Times* reported news received the previous day by telegram from Clyde about 'an unprecedented fall of snow all over the district since 2 a.m. on Wednesday 8 July'. Four days later, it reprinted a piece taken from the 10 July issue of the *Dunstan Times* that provided additional detail: six inches of snow had fallen between 2 a.m. and 5 p.m., with the likelihood of a hard frost to follow. Most South Island newspapers reported severe frosts, snowstorms and flooding in the larger rivers throughout July 1868, with a continuance of squally weather into August. In Dunedin, the heaviest snowfalls so far that winter were on 3 August, and there were falls of snow and sleet the following day: 'The severity of the weather prevailing [in Dunedin] during the last few weeks has affected trade.' The Central Otago town of St Bathans had experienced six weeks of continuous frost, and on 5 August the newspaper reported, 'The weather has been unusually severe [in the Otago goldfields] and has interfered much with working in the higher districts. Snow has fallen heavily almost all over the Province. This . . . gives promise of an abundant supply of water for the spring and summer.' In its 6 August 1867 edition, *The Press* printed a short piece about widespread storm damage in the northern districts of Waipara and Cheviot, where 'Travelling has been rendered almost impossible owing to the quantity of snow that has accumulated on the downs. Some of the sheep farmers have suffered very heavy losses, as the snow has unfortunately fallen during their early lambing, inflicting great injury on their flocks.' The weather in August remained cold and squally throughout Otago and into Canterbury.

There were few such storms during the 1880s, but a prolonged, unseasonal cold snap during September 1884 was especially severe in the upper Waitaki Valley as well as nearby areas in Otago and the Mackenzie Country, and had an adverse impact on young lambs, as William Shirres reported to his financial backers:

> Since my last letter [3 September 1884] we have had a very severe storm. Indeed, the worst we have had since the commencement of winter. Fortunately for us, our lambing had scarcely begun. It [the snowstorm] commenced with a heavy rain and ended up with snow during the night, and during the three succeeding days there was an almost incessant cold rain . . . most of the neighbours will have suffered very much as their lambings were nearly over, and so cold was it that many lambs over two weeks old died. This storm, occurring as it did over the South Island at the time of the general lambing will without doubt affect the surplus of this country compared with last year, and probably have an influence on the export of frozen meat . . . . I was at the Otematata Station yesterday and the manager there told me he would be pleased if he had 5,000 lambs at marking time. The lambs were lying dead on the run in hundreds.[11]

Rural people learned to forecast a simple change in the weather, but predicting the onset of a major snowstorm, severe cold or a protracted spell of heavy rain was beyond their ability. An episode of adverse weather was not an annual event but could be a major setback to the operation when it did occur, especially in conjunction with fog, drought or a prolonged frosty spell. There was worse to come in the 1890s, and on 2 July 1891 the manager of Ida Valley Station wrote to his landholder:

The snow came on June 6th – four days after I [last] wrote you. It was not deep – only about 3 inches – but a high and bitter wind sent it into drifts, and as there have been hard frosts ever since, it gives way very slowly . . . . [The sheep on the upper ground] I am sure suffered from want of water (to drink in the autumn). June has been a trying month for them: first the snow, and since then terribly hard frosts and a great deal of fog. The fog I think must be worse for them than the snow, as every tussock is as hard as a board all day long, and never a blink of sun! . . . . There is only a dribble [of water] in the creeks, but of course this is all the better as it means that what little snow there was has gone into the ground. We want snow very badly for the [good of the] county.[12]

Four years later, on 1 June 1895, he wrote:

We have had a heavy fall of snow – from 9 to 12 inches all over. It started last Saturday morning and continued all Sunday, and the fall was followed by the most intense frosts I can remember. The cold both outside [the house] and in was terrible. The first thaw commenced today, and it is not unlikely [it] will be followed by another fall [of snow] . . . . I don't remember such severe weather so early in winter, and in the present absence of feed it will be a very serious matter if there is much more of it [i.e. extreme cold].

This was the first in a series of severe storms that year, and on 1 August 1895 the manager of Ida Valley Station advised the absentee landholder:

Since I wrote you on 11 July we have had constant and intense frosts, with more snow at the end of the month. Very little of the snow has gone since I wrote so that during June and July the ground has practically been covered. I don't think that on this run more than the total average of snow for May, June and July has fallen, but never in my time has it ever lain anything like so long. For nine weeks the sheep have had nothing to eat but tussock (there's not much more anyhow) [and] . . . . most of them look better than they have any right to look . . . . You will have read and heard of the appalling state of things on some of the Stations up here – and the Waitaki Country seems worse – Weir was down Wedderburn way last week, and his account was pitiable in the extreme. I am afraid almost all the sheep on the Hawkdun Range must be wiped out. Even those that were brought down with the greatest difficulty to the lower country must succumb; their wool, and even the case [i.e. wool, skin and underlying tissues] of some has been eaten off by each other, and they are dying on the roadside by hundreds and being skinned at the rate of 500 a day! A Blackstone shepherd who was here last night said he was sure that full 5000 of the sheep (on the Hawkdun Range) were dead at the camp before they got the live ones started. Morven Hills I hear (and believe) won't save a single hogget.

The adverse economic impact of the 1895 snowstorm can be gauged from Preston's outlays on sheep bought during 1896 to replace those lost during that storm. For Haldon Station, he bought 130 rams, 3000 ewes and 3400 wethers for an outlay of £1645; for Ben Ohau Station, he bought 7383 sheep for £1296; and for Aviemore Station, he bought 5301 sheep for £1265. These replacement animals were sourced from back country properties across southern New Zealand, many of them near Lake Wanaka in Central Otago. He also applied for rates relief on account of the exceptional losses of livestock – several horses, ten of his 52 cattle, and 9000 of his 14,000 sheep – at Ben Ohau Station.[13] The combination of high rentals, the heavy financial burden of rabbit control and severe winter stock losses brought about by snowstorms in 1895 forced him to relinquish the station in 1897.

Conditions remained extreme for most of the remaining years of that decade, as Morton informed Preston on 9 December 1897: 'The weather has been the coldest I have seen for this time of year. Last Tuesday, 30th November, was the coldest day I have been out in for a long time. It was our first day's mustering and we were stuck behind rocks for about 3 hours with a hail storm.'[14] On 12 July the following winter he wrote, 'We have had very rough weather since you were here. There has been more rain fallen in [indecipherable] some time than there has been since I came to Haldon, and a good deal of snow. There was a slight fall [of snow] at the house, but not much, but on hill tops there has been a good deal . . . . There was 18 inches of snow at the [Burkes] Pass, and 6 inches at Fairlie . . . . Now about buying Haldon back again. I would like to, but am afraid to put everything in it for fear I should get another wipe out with snow, and hence to leave me with nothing.' On 9 August he told Preston, 'We have had some bad weather since I wrote you last. We had 7 or 8 inches of snow at [the] house and [it] lay a long time on account of the short days and hard frosts.' But the misery continued, and on 27 September he sent a long and detailed report to Preston:

> We have had dreadful weather for the last month back: snow, sleet and rain every few days. There were about 4 inches of snow at house and about 18 inches at Burkes Pass. There was a good coating on Big Ben Range and Back Blocks. It wound up with half a day's thunder on Friday last and came on to rain about 4 p.m., and rained heavy all night, a warm rain. The next morning there was water everywhere, and Stony Creek was very high. There was about a foot of water running between house and wool shed. I had a job to save the dam. It swept away the scrub work [placed to trap large and small stones] at [indecipherable] crossing of new fence on Stony Creek line. Some of the [fence] posts [were] laid over [on] two sides of new field around oat crop. It has carried away all the scrub work, wire and all across the crossings in Ross's Creek and Stony Creek Gorge. There has not been so much water about since the thaw of snow in '95. Weather has been fine this last week . . . . the creeks are [now] so high it is impossible to get sheep across, and they keep high [as] the warm days keep melting the snow.

The appalling weather continued into October, when the ground around Haldon homestead remained covered in snow: 'We have had rather rough weather lately. We had a slight fall of snow, about 4 or 5 inches about three weeks ago, and it is still lying. There has been no snow [since then] and been very cold – about 18 degrees of frost. Some of the time freezing all day.'

AT HOME IN A CHALLENGING ENVIRONMENT

Not all adverse weather systems caused significant problems for farm and station holders during the second half of the nineteenth century, and several years could pass between two of them in succession, but my reading of settlers' diaries and letter books showed they were aware that different parts of southern New Zealand experienced particular mixtures of environmental, economic and social problems driven by the weather. To evaluate that, I divided the area into six regions to ensure the best possible coverage for my documentary sources:

mid- and south Canterbury, the middle and upper Waitaki Valley including the Mackenzie Country, Central Otago and the Maniototo, the Lake Wakatipu basin, northern Southland and south Otago. Then, from the more than 50 farm and station diaries available to me, I extracted records of

i.   prolonged snow and severe cold causing stock losses;
ii. sufficiently dry conditions for a major reduction in plant growth and an adverse effect on the well-being of people and livestock; and
iii. heavy rain on the property – or rain and snowmelt in the distant head-waters of large rivers – with flooding and significant environmental damage.[15]

The findings are shown in both parts of Figure 3.1.

All six regions experienced problems with snow, severe cold and either too much or too little rainfall, but major environmental problems triggered by low rainfall were less evident in the Lake Wakatipu basin, south Otago and Southland, and episodes of flooding were less frequent in the upper Waitaki Valley and Mackenzie Country, Central Otago and the Maniototo. The greater incidence of flooding in Canterbury was a consequence of that area's large rivers receiving water from melting snow and heavy rain in their alpine catch-ments during northwest storms. If rain-bearing winds from the southwest or the easterly quadrant did not eventuate, then widespread drought could be experienced in spring and summer on the plains even when large snow-fed rivers were in flood. The complex seasonal interplay of weather systems and topography in southern New Zealand was recognised by settlers, and John Wither frequently recorded that heavy rain and snowmelt in the headwaters of Lake Wakatipu had preceded flooding in low-lying parts of his property by a day or more. Across southern New Zealand, settlers were also aware of regional differences in weather and climate, and understood that adverse weather could be experienced widely (for example, the heavy snowfalls and floods of 1867 and 1868), or be severe in one part of the district and of little consequence in another (the February 1868 storm that swept down the east coast of the South Island to a little beyond Dunedin, but not as far as Invercargill, is an example of this). Although landholders did not have access to diagrammatic summaries like those in Figure 3.1, their diary entries showed a growing awareness of geo-graphical and historical variations in weather conditions.

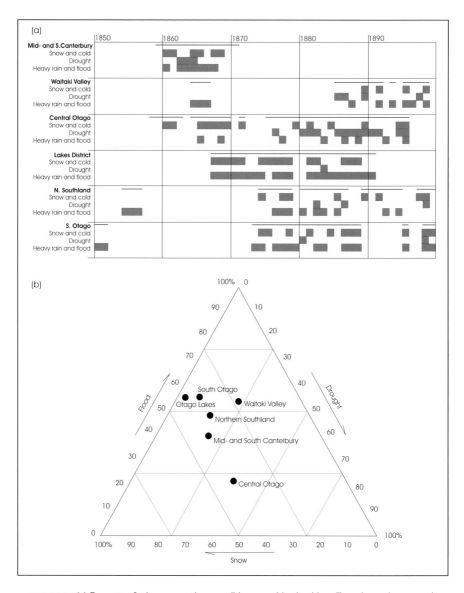

FIGURE 3.1 (a) Reports of adverse weather conditions resulting in either (i) prolonged snow and intense cold, or (ii) critically dry soils, or (iii) heavy rain and flooding in streams and large rivers between 1850 and 1899; and (b) a triangle diagram to show relative frequencies of those three adverse weather and weather-related conditions for each of six regions in southern New Zealand. In (a), thickened lines show the years covered by the documentary sources for a region and a shaded block marks at least one annual mention of the named condition. SOURCE: SEE NOTE 15

The belief in a widespread, benign, Mediterranean-type climate foundered on the experiences of the first generation of European settlers in southern New Zealand. The first ten years were comparatively mild, but during the decade beginning in the early 1860s settlers experienced major floods, severe gales, extended periods of very low temperatures and heavy snowfalls. Their vulnerability to those conditions was enhanced by flimsy dwellings, inadequate fencing and a nascent infrastructure. The weather events that caused occasionally severe disruption could not have been foretold, and settlers were unaware that what was happening elsewhere in the country would likely soon affect them: in fact, in the first two decades of organised settlement rural people knew little of how other settlers and areas were faring from day to day. It was, however, important for them to experience such events, observe the problems they caused, make appropriate changes to their property lay-outs, reduce stocking densities on vulnerable land, and regulate flows of water into and through their properties. They learned from occurrences of unusually severe weather, just as they did from the normal range of weather conditions, but while they could insure against damage from recurrence of an extreme event by revising their land management routines, they could not foretell its occurrence or its severity.

CHAPTER FOUR

Away with the Old
What Place for Native Plants and Animals?

In 1879, barely three decades after the start of organised settlement in Canterbury, a local resident, J. C. Firth, published an article in a widely read farming journal on the destruction of native forest resources:

> We are destroying our noble forests, the growth of centuries, and to this day (except by Mr Vogel's Forests Act) we have not made even an effort to stay the reckless tide of havoc and ruin which is sweeping them away. We are neither conserving the old nor creating new. If a squatter, having come into possession of a rare, valuable, and unrivalled flock of sheep, and, without taking a single step to perpetuate the famous breed, should proceed to boil down the entire flock, young and old, he would be denounced as a lunatic or a public enemy, or both. Yet we are now busily engaged in perpetrating a similar or worse enormity.[1]

At the same time, S. Grant and J. S. Foster, who were visiting the country on behalf of other tenant farmers in Lincolnshire, had noted that land held by Māori near Whanganui was available for lease but that the leasee was bound 'to leave a certain amount of timber on the run'. They also observed the destructive behaviour of Pākehā settlers: 'The varieties of wood [in the Seventy Mile Bush, Wairarapa] are very numerous and would serve for every purpose for which wood is used, if a market could be found for it.... [Without investment in infrastructure, sawmills and market development, it is impossible to sell large quantities of native timber at a profit, so] when a forest is being cleared the wood is all burnt, either standing or after being felled, and the most beautiful timber may be seen burning on the ground, literally in heaps, as if it were so much rubbish.'[2] Had they visited forested areas on Banks Peninsula, as well as along the eastern foothills of the Southern Alps, the Catlins or western Southland, they would certainly have reported similar wastage. Although they travelled widely in New Zealand, they did not comment on the draining of large and small wetlands or the burning of tracts of tussock and low shrubland.

Despite their destruction of much of the natural landscape they moved into, settlers initially depended on tussock communities to feed and shelter their sheep and cattle; exploited large and small stands of native bush – remnants of once more extensive forests that had thrived a millennium earlier – for lumber, fencing materials, fuel, food, decorative plants and recreation; and harvested reeds for thatching as well as for the bases and caps of their hayricks and stacks of harvested grain ready for threshing in winter. Compared with the numerous, often detailed, entries in farm and station diaries about daily weather conditions, major floods, prolonged episodes of drought, and problems with pest plants and animals, I found that entries about native plants, animals and ecosystems were infrequent, even cursory. This chapter draws primarily on more than 50 runs of farm and station diaries, and relates the information obtained to the views of commentators who were even then warning rural people about the economic consequences of their environmentally and ecologically disruptive activities.

A RESOURCE FOR THE TAKING

Names of native plant and animal species were occasionally noted in settlers' diaries, and Table 4.1 lists those recorded by Joseph Munnings between 1859 and 1866.[3] He saw them in Christchurch, on Banks Peninsula, across the plains of mid-Canterbury and in the eastern foothills of the Southern Alps, an area that Charles Torlesse had surveyed in the late 1840s and 1850s.[4]

TABLE 4.1 Number of references to native plants and animals in Joseph Munnings' diaries, 1859–1866, listed in decreasing frequency of citation.

Number of references	Native plant or animal
Identifiable native plants	
58	Black pine [matai, *Prumnopitys taxifolia*]
44	Bracken / fern [*Pteridium esculentum*]
42	White wood from Hoon Hay [probably kahikatea, *Dacrycarpus dacrydiodes*]
28	Manuka [probably kānuka, *Kunzea ericoides*]
14	Toot / tutu [*Coriaria* spp.]
3	Black birch [probably black beech, *Nothofagus solandri* var. *solandri*]
2	Flax [harakeke, *Phormium tenax*]
2	Fuicha [fuchsia, *Fuchsia* spp.]
2	Palm / tea palm / tī palm [cabbage tree, tī, *Cordyline australis*]
2	Supplejack [*Ripogonum scandens*]
1	Birch [probably beech, *Nothofagus* spp.]
1	Cress [possibly *Lepidium sativum*]
1	Totara / totra / tutra [*Podocarpus totara*]
1	Decorative grass [probably flowering shoots of toetoe, *Cortaderia splendens*]
1	Nigger head [*Carex secta*]
1	White birch [probably beech, *Nothofagus* spp.]
Unidentifiable native plants	
139	Sawn construction timber
56	Fence posts and rails
53	Firewood
19	Palings
8	Shingles
2	Thatching
2	Unnamed decorative plants
1	Forest ferns
1	[Tussock] grass
Native animals	
3	Duck / grey duck
1	Dogfish
1	Rock cod

SOURCE: SEE NOTE 3

Through his carting business, Munnings met the Phillips family of Rockwood Station. Three of the boys managed The Point Station nearby,[5] and their diaries contained occasional references to native plants and animals (Table 4.2).

TABLE 4.2 Number of references to native plants and animals in the diaries of The Point Station in mid-Canterbury, 1866–71: entries in each group listed in decreasing frequency of citation.

Number of references	Native plant or animal
Plants	
71	Tussocks
61	Firewood from bush
29	Fence posts, rails and standards
11	Decorative bush plants
4	Brushwood for hayricks
4	Flax [probably harakeke, *Phormium tenax*]
1	Irishman [matagouri, *Discaria toumatou*]
1	Rata [*Metrosideros* sp.]
Animals	
8	Eel
5	Weka
4	Paradise duck
4	Unnamed birds' eggs
3	Gulls' eggs
2	Lark
2	Terns' eggs
1	Pigeon
1	Quailhawk
1	Swamp hen [pūkeko]

SOURCE: SEE NOTE 5.

From January 1858 to the close of 1863, the Macdonald family of Orari Station, south Canterbury, relied on nearby stands of bush for fencing materials, fuel, lumber and shingles for the roofs of their living quarters (Table 4.3).[6] They were also preferred places for family picnics in summer. A few large woody plants were named in the family's diaries, notably black pine (matai), mānuka (probably what we know as kānuka) and white pine (kahikatea). What they called manuka poles were used for fence rails and hurdles, as well as for placing across bush tracks to prevent loaded wagons from getting bogged in the muddy soil. Mature cabbage tree stems were harvested for the same purpose, and their green shoots – like the stiff leaves of flax and toetoe – were used to thatch haystacks and farm buildings. The word 'tussock' did not appear in these diaries, but there were references to 'grass fires' in late winter and early spring. Wood pigeon was the only named native animal, and it was shot for sport and the pot.

TABLE 4.3 Numbers of dray-loads of wood taken annually from nearby areas of native bush for use on William Macdonald's property, Orari Station in south Canterbury, and grouped by the following uses: fencing materials (hurdles, palings, posts and rails), firewood (for domestic purposes and heating water for washing wool), lumber (building construction, including laths, piles and shingles), and for laying across muddy bush tracks to facilitate removal of fuel and sawn lumber.

Year	Fencing materials	Firewood	Lumber	Shingles	Laying on muddy tracks
1858	18	36	19	5	16
1859	53	40	13	1	9
1860	43	40	13	3	13
1861	35	106	36	5	7
1862	38	80	3	-	-
1863	69	113	11	-	-

SOURCE: SEE NOTE 6

The South Canterbury Museum holds a run of diaries from Grampian Hills Station in the Mackenzie Country from 10 October 1862, when a pioneer family settled the property, until 5 March 1866.[7] The day after the family set up camp in their new home, a dray was dispatched to a distant stand of native bush for firewood. Most weeks thereafter, it was sent out to collect wood for fuel, construction materials and fence posts, or to ferry firewood to outlying shepherds' camps. Family members and hired labour harvested thatch from reed beds, carted it to the cottage and out-buildings then under construction, and brought back loads of 'scrub' – most likely matagouri – for fuel at shearing time.

The diaries of Kate Sheath provide another early account of pioneering in south Canterbury.[8] She and her family lived in the downlands and low hill country beside the Little Opawa River near Albury. Her diaries, which ran from December 1862 to June 1868, recorded the success of menfolk in shooting native pigeons, ducks and kākā; summer picnics in cool stands of native forest; forays into the bush for tree ferns, decorative shrubs and vines for transplanting around the homestead; and winter trips to large remnant stands of forest in the hills for dray-loads of fuel and fence posts. Like her contemporaries, Kate Sheath kept kākā and other native birds as household pets.[9]

For people living far from stands of native forest in the hill country and interior basins of southern New Zealand, the charred trunks of forest trees preserved in moist topsoil – commonly referred to as 'mountain timber' – were valued. In their diary entry for 5 September 1871, the Teschemaker brothers on Otaio Station in south Canterbury recorded that sufficient fossilised wood had been snigged out of mature tussock grassland by hired labourers for 1835 stakes, 48 strainers and 41 fence posts.[10]

There were similar entries in diaries from nearby Raincliff Station.[11] The available diaries for this property cover four consecutive years during the 1860s and 1870s followed by a gap until 1896, by which time Raincliff Station was a mature operation. The property was then almost completely divided into fenced paddocks and fields; homestead and out-buildings had been constructed; plantings of introduced tree species were already providing fuel and sawn wood; arable land was under cultivation or had been planted in crops and improved pastures; and introduced plants were deployed for shelter, river and stream control, decoration, and food for livestock and people. Taken as a set, these diaries show the changing importance of native plants during the second half of the nineteenth century: native species accounted for at least half of all entries mentioning plants in the first four years (Table 4.4), but by 1896 this component had fallen to 35 per cent. Raincliff Station was not alone in this regard. Despite the sizeable number of native species suitable for shelter in southern New Zealand, few were used on this property. The short-term advantages of introduced tree and shrub species for hedging and shelter were well known but few landholders appreciated their long-term cost, despite published statements such as this: 'The indigenous shrubs are slower [growing] than pines or eucalypts or wattles. This at first places them at a disadvantage, but afterwards becomes one of their greatest merits. The rapid-growing exotics serve us with little cost for five or six years and then begin to make us pay heavily for their benefits.'[12]

TABLE 4.4 References to useful native and introduced plants at Raincliff Station, south Canterbury, expressed as percentages of respective annual totals.

	1868	1869	1870	1871	1896
Native plants					
For construction lumber	12	11	22	13	1
For fencing materials	15	19	13	3	20
For fuel	16	17	20	30	14
For planting around homestead	4	1	0	4	0
For other uses	1	11	6	1	0
Introduced plants					
Fruits and vegetables	22	15	19	11	7
Crop and pasture plants	2	3	11	15	28
Shelter, hedges and decoration	25	14	9	16	29
Weeds	3	9	4	7	1

SOURCE: SEE NOTE 11

Between the late 1870s and early 1880s the only native plants named in the ledger books for Te Waimate Station in south Canterbury were black pine (matai), which was valued for fence posts and rails, and flax for thatching hay-ricks and stacks of wheat held on the property until the grain could be threshed later in the year. Nevertheless, the ledgers show that many loads of fence posts and lumber were being brought onto the property from sawmills in the nearby Waimate Bush and from stands of valley forest on the eastern flanks of the Hunters Hills.[13] On this large property, like farms and stations elsewhere in southern New Zealand, the economic, environmental and cultural values of native plant and animal species were not appreciated even though they effectively subsidised and supported transformation of the landscape by the first generation of European settlers.

James Preston, who leased several large sheep stations in the Maniototo, the Waitaki Valley and the Mackenzie Country at different times during the late nineteenth and early twentieth centuries, kept diaries for each of them during his periods of residence. His diaries for Haldon Station in the Mackenzie Country, albeit with occasional long and short gaps, covered the period from January 1878 to September 1891 and contained numerous references to native plant and animal species, notably pūkeko and 'parroquets' (possibly kākā), as well as plants from the wetland, tussock grass, low shrub and forest ecosystems on his property. He mentioned beech by name – it had almost certainly been brought in as sawn lumber from western Southland – cabbage tree, mānuka, snowgrass for thatching farm buildings and haystacks as well as for stuffing mattresses, an unnamed 'shrub' (possibly matagouri), toetoe, tōtara, tussock and tutu. He occasionally identified native species used for fencing, and in his diary on 30 May 1904 recorded the cost of erecting a rabbit-proof boundary fence: £25 5s for 505 kōwhai fence posts, £4 for 16 kōwhai strainers and £5 14s for beech fence posts. Kōwhai is one of New Zealand's more durable native woods, and during the nineteenth century 'was tried for sleepers, house blocks, fence posts and piles. It soon built up a reputation for durability but the small size of the tree restricted its use.'[14]

Several shrubby native plants were suitable for hedges in the lowlands, and amongst the thirteen species listed by agronomist and plant scientist A. E. Green were *Griselinia littoralis*, *Myrsine urvillei* (probably *Myrsine australis*), *Olearia ilicifolia*, *Olearia dentata* (probably *Olearia macrodonta*), *Pittosporum euge-nioides* and *Pittosporum tenuifolium*. All grew in southern New Zealand, and the first named species was especially effective when closely trimmed after flowering.[15] Green discussed the selection, propagation and maintenance of native hedge plants and argued that they should be established on a farm for their utility and beauty as well as a way to replenish 'what is gradually becoming extinguished: the New Zealand bush'.[16]

Like his nineteenth-century contemporaries, John Wither of Sunnyside Station in the Lake Wakatipu basin occasionally recorded names of native plants and animals in his diaries. Most were organisms with novelty or short-term value to Wither and his family, and were later displaced by introduced plants and animals. He and his employees collected bunches of tussock to stuff mattresses for shearing gangs, and cut bundles of snowgrass shoots to make brooms for use in the shearing shed. Mature tussocks provided shelter from cold winds as well as herbage for pregnant ewes trapped by heavy snow on higher ground. Patches of valley forest on the property or in nearby hill country provided wood for building piles, fence posts and construction, as well as fuel for cooking, heating the house and boiling water to scour wool. Most other native plants were essentially candidates for clearance.

Like tussock growing on cultivable land, bracken ('fern' in Wither's diaries) was first topped and then its underground tissues were grubbed out, left to dry and burned. Expanses of bracken were replaced by sown pastures of cocksfoot, ryegrass, and red and white clover for grazing by sheep and cattle. What Wither termed 'thorn' was most likely matagouri, which he and his employees grubbed out and burned on-site during winter. As elsewhere in the lowlands of southern New Zealand, tutu, which Wither termed 'toot', was a major nuisance because sheep and cattle that ate its young shoots and ripe berries could be poisoned and die within a day unless treated. Before the end of 1882 there were ten references to this plant in Wither's diary, most of them in late spring and early summer. Another native plant that caused difficulties was bush lawyer, which grew in thick masses in the bushy vegetation of better-watered, sheltered valley sites where a woolly sheep could be entangled by the hooks on its leaves and long wiry shoots.

In his diaries, Wither referred only once to the native eel, three times to what were presumably native flies, four times to shags and 32 times to kea. The latter were vigorously hunted because the bird was known to attack live sheep: 'A number [of sheep] killed with keas at the George . . . . Saw 10 sheep killed with keas on Kamper' (5 August 1874); '8 sheep killed with keas on Kamper' (6 August 1874); 'Put the sheep onto the George Spur; some killed [when they fell] over rocks, some killed by keas' (26 June 1875); 'Went round the back of Bayonet Peak. Saw one sheep killed by the keas' (24 July 1875); and 'Saw one sheep killed by keas' (2 May 1876). In 1872 and 1873 the prominent Canterbury runholder and amateur naturalist Thomas Potts published two articles in the British science magazine, *Nature*,[17] in which he documented a shift in food preference from herbivory, supplemented by insectivory, to carnivory in this species of parrot, a topic that continued to interest naturalists well into the twentieth century.[18] Wither also reported shooting shags – presumably the black shag of inland water bodies that was believed to predate on introduced trout – and eating their eggs. On 27 January 1882 he recorded having been paid a bounty of 2 shillings for each dead bird.

Wither evidently viewed his large property as an ensemble of native ecosystems that he had to transform with fire, felling and cultivation, aided by introduced plant and animal species, to ensure an economically productive and congenial environment for himself, his wife and their family. To that end, he brought in decorative and pasture plants, vegetables, and fruit-bearing shrubs and trees to transform the appearance of the land and the species composition of its semi-natural and managed ecosystems, thereby creating his own, and to him more pleasant, version of nature. Nowhere in his diaries is there any expression of delight in native plants and animals, nor expression of regret at their eradication. Wither was literate, a Scot, a devout Christian and a good parent, but if he had any deeper feelings about the landscapes and ecosystems that he and his employees were so single-mindedly transforming, then he did not record them in his diaries.

John Finlay, who worked a small farm at Makikihi in coastal south Canterbury, wrote in his diaries about drawing out lengths of charred forest tree wood, which he used for fencing, from tussock grassland on the Hunters Hills.[19] He also made frequent trips to Ambrose Jackson's bush on the eastern flanks of the Hunters Hills, where he bought sawn native timber for fencing and firewood. He cleared cabbage trees from the grassy downlands in preparation for ploughing, staked garden peas with mānuka twigs, used the shoots of native shrubs for thatching out-buildings, and transplanted tree ferns and ornamental flax plants to garden beds around his house. Almost certainly, Finlay did not realise that some of the groves of cabbage trees on his farm could have been planted by Māori several decades earlier for later harvesting as a source of fructose: the sugar was produced when the mature stems and large below-ground tissues were cooked in an umu.

For settlers like David Bryce and his family on their farm at Lovells Flat, midway between Milton and Balclutha in east Otago, any remaining areas of swampy land were little more than sources of fossilised wood for fence posts, and rushes for thatching stacks of hay, oats and wheat. At the same time, recent settlers like the Cody family at Lime Hills in northern Southland made extensive use of native plants.[20] Their diaries for the years between 1904 and early 1916 contain 41 references to native species and their uses: forest and forest-edge plants like kōwhai, kahikatea and unnamed woody plants were felled for fence posts, droppers and rails, as well as for construction timber and fuel; the stiff leaves of wetland plants such as flax were plaited into ropes; rushes were collected for stack bottoms and thatching; tussock shoots were used for thatching haystacks; and small shrubs were collected for household fuel. After a long gap, the next extant diary was for 1949. By then, the farm was experiencing more frequent short-duration floods than three decades earlier, thistles were the only named weed, and native plants and animals were not mentioned. The phase of environmental transformation initiated on the Cody family farm at the start of the twentieth century was largely complete.

Homestead, Holme Station, south Canterbury, 1870s, showing a mix of tussocks, cabbage trees, gorse hedges and eucalyptus trees. S.34, SOUTH CANTERBURY MUSEUM, TIMARU

Throughout southern New Zealand, landholders, tenants and managers responded in various ways to average and exceptional environmental conditions. How Wither thought his property on the shores of Lake Wakatipu should be developed differed from the views of two contemporaries, the successive Scottish managers of Ida Valley Station in Central Otago, during the final three decades of the nineteenth century. On 19 August 1875 the manager wrote to the absentee landholder, 'I have previously over-estimated the carrying capacity of this Run . . . [On nearby] Galloway Run . . . there is absolutely no feed whatever on thousands of acres, and as far as I can judge little chance of there ever being any permanent improvement in the country without total rest for a season at least.' Entries in the diaries for Kauru Hill farm in north Otago and a pastoral property in the middle Waitaki Valley, both of them operated by the same family, tell a similar story of ecological savvy: for example, 'Mathew and I went to Otepopo Bush, having sent a cart there to collect shrubs' (1 September 1875), and the following day 'John and his mates planting out shrubs near stable.'[21] They were not alone in these respects, but most landholders felt that the landscapes of their properties had to be comprehensively transformed to demonstrate their abilities and ensure their economic well-being.

A RESOURCE TO EXPLOIT OR CONSERVE?

In an early reference to native species as pests on recently settled land in Southland, James Menzies recorded in his diary on 9 July 1855 that 'kites are killing newly dropped lambs, not content with the dead ones'.[22] Seven years later, Joseph Munnings, who had arrived in Canterbury in 1859, wrote about a trip that he and a companion had made to Akaroa on Banks Peninsula:[23]

> We arrived in Pigeon Bay about eleven [and] we went on shore with the boat . . . .
> We went some distance up the Akaroa track and out and in the Bush. W Packard shot several small birds; some very fine trees. The Bush fast receding, the hand of man is lying in wait and turning the wild Bush into green pastures. It is expensive clearing it. Cut the best of the wood for timber and firewood, and then set fire to the underwood. Stub a hole here and there all over it and put in potatoes, which grow very fast and fine.

A handbook published for the guidance of intending settlers described the landscape of lowland Canterbury and outlined a model for how it might be transformed: 'The plains can never be made to look pretty as a landscape; they may, however, be rendered less monotonous in appearance by enclosures, the growth of hedges, and particularly the planting of trees.'[24] The following year, in its 3 January edition, *The Press* corrected a statement printed the previous week about the herbage a herd of cattle had been grazing: 'The beef was [in fact] fattened on native grasses on Mr Moses Cryer's run near the mouth of the Rakaia [River].' Four years later, on 9 October 1867, the *Bruce Herald* published an apparently original article under the headline 'Native Grasses', in which the author posed the key question:

Are our farmers and graziers correct in assuming as a determined fact that 'English Grasses' are the best or only seeds which should be used to form a permanent sward of pasture? We take leave to be very skeptical on the matter.... we are convinced some of the grasses composing the native sward must possess highly nutritious and fattening properties.... we cannot see any reason why the nature, [growth] habits and properties of native grasses should not be made the subject of extensive experiment and trial involving plots of ground equal in quality and the same size.... To make the experiment of really extensive value, it should be simultaneously carried out with English grasses of a somewhat similar variety.... [He then argued for this work to be done in the Dunedin Botanical Garden.] We would seek very strongly to excite a spirit of investigation and enterprise amongst our farmers who have hitherto been too slavish in adopting cultural customs and practices without knowing the 'why' and the 'wherefore'.

These three opinions – the first holding that native species can compromise the economy of a rural property, the second seeing native species standing in the path of progress, and the third envisaging realisable opportunities for native species – encapsulate the ambivalence of settler society towards native plants and animals. The North American environmental historian William Cronon has argued that what we think of as wilderness should not be held apart from the world of people, but seen as an area full of symbols that give value to humanity.[25] In southern New Zealand, as in frontier lands of the United States of America, wilderness areas were initially viewed as waste lands awaiting transformation by people. In both countries, that stance began to change towards the end of the nineteenth century, and this shows in how settlers treated, and what they thought about, native species.

A striking feature of the informal documentary record from the first two decades of organised settlement in southern New Zealand is the infrequency of references in farm and station diaries to problems associated with introduced animals and plants. This was not the case with native species, however. Among endemic animals regarded as problematical were a native parrot, kea, which had been found to attack sheep and expose the animals to infection – a bounty was paid for the bills of dead birds; and pūkeko, because of its habit of pulling up emerging crop plants. Most native birds were innocuous, and few of those shot by members of Kate Sheath's family in south Canterbury were commonly acknowledged pests. Relatively more native plant species were viewed as weeds, and Leonard Cockayne listed several that grew in the tussock grasslands: notably, mat raoulia, mountain twitch, pungent heath, red piripiri, spineless piripiri, swamp lily and turf coprosma.[26] From the early years of organised settlement, the poisonous shrub tutu was universally viewed as a serious nuisance, while prickly woody plants that grew in damp gullies and along the bush edge often proved a danger to sheep. On 5 August 1884 William Shirres wrote to the financial backer of Aviemore Station:

> In most of the gullies on the run there is a great deal of scrub. Some of it [is] over twenty feet high, and impenetrable to everything except [word indecipherable] .... A number of sheep get lost in the scrub, and old sheep do not care to enter these gullies, especially with their lambs. Lately, I have had the men out burning whenever there was a chance, and have been out burning myself. I have succeeded in getting [rid of] a great deal of it and in the ashes I have sowed rye grass and clover seed, which I expect will improve the run very much.[27]

However, several of the plants that were regarded as nuisances in some circumstances were seen as beneficial in others. For example, matagouri was treated as a weed when it impeded access to grazing, but at shearing time was collected as a fuel for heating water to scour the wool clip. Bush lawyer sometimes created an impenetrable tangle that could be lethal to sheep, but was also valued because bundles of its shoots formed a sturdy mesh that could trap rocks and cobbles in stream beds at risk of erosion during times of high stream discharge, or strengthen low earth dams erected across small streams to pond water needed for scouring the wool clip. Dry fronds of bracken fern were used to stuff mattresses and as bedding for livestock even while fernland was being torched in readiness for ploughing and sowing in pasture plants, grain or root crops. Tussocks were grubbed out to clear the area for crops and sown pasture, yet afforded shelter for livestock during storms and provided browse throughout the year. Rushes and other wetland plants were burned to create open ground for sowing pasture and crop plants, but were also sources of stiff shoots to form the foundation for a hayrick or to thatch grain stacks and farm buildings. Even tracts of native forest were concurrently perceived as impediments to progress and vaults of useful biological resources.

Two four-horse ploughs, Waimate, south Canterbury.
2002-1026-00052, WAIMATE HISTORICAL SOCIETY AND MUSEUM

Early generations of European settlers were not insensitive to the physical appearance of the indigenous landscape. As John Finlay wrote in his diary on 11 February 1881, 'the banks of the Makikihi river [leading to Ambrose Jackson's bush] are here clothed with luxuriant bush of great beauty, and the drive for the last two miles or so being one of the prettiest to be found anywhere in the neighbourhood'.[28] In 1898, however, William Pember Reeves published an anthology of poems, including his well-known and frequently cited 'The Passing of the Forest', a paean to a lost world that included these bleak words: 'Bitter the thought: Is this the price we pay – The price for progress – beauty swept away?'[29] A decade later he expanded this theme in a book published during his term as High Commissioner for New Zealand in London:

And then there is the great area deliberately cut and burned to make way for grass. Here the defender of tree-life is faced with a more difficult problem. The men who are doing the melancholy work of destruction are doing also the work of colonisation.... They are acting lawfully and in good faith. Yet the result is the hewing down and sweeping away of beauty.... Not that my countrymen are more blind to beauty than other colonists from Europe. It is mere accident which has laid upon them the burden of having ruined more natural beauty in the last half-century than have other pioneers.[30]

It was not just the natural beauty of southern New Zealand that was at risk. The area's economic potential was being compromised by large-scale vegetation clearance. From the 1860s onwards the eminent naturalists J. F. Armstrong and W. T. L. Travers documented the increasingly parlous status of native plants and animals in Canterbury. Armstrong recognised the economic worth of kahikatea, matai, rimu and tōtara, but believed that they lacked the durability of wood then being imported from Australia.[31] He saw black beech as a useful source of tannin for preserving leather. For his part, J. B. Armstrong did not doubt that the native flax would become a valued industrial resource, and perceived an important, long-term role for native grasses in the rural economy:

Not until the feeding value of native grasses, more especially in comparison with imported ones, has been ascertained, and their applicability or inapplicability, as the case may be, to ordinary meadow and pasture purposes has been determined by actual experiment, I cannot but think that their indiscriminate destruction is an act of folly. All stockmen agree in praising the feeding qualities of native grasses, and I have personally seen sufficient to satisfy me that many of them could most advantageously be mixed with imported grasses in forming artificial pastures.[32]

He was not the only landholder or observer to take this stance. In October 1872 the manager of Levels Station wrote to the absentee landholder, 'The native grass however looks well enough over the whole of the different runs with plenty of feed but somewhat dry in places.'[33] He had previously reported burning sections of the property and sowing newly ploughed land in English grasses, which did not do so well under drought conditions. In 1879, a Christchurch businessman and officer in the Canterbury Agricultural and Pastoral Association,

Robert Wilkin, described an early attempt to evaluate the potential of native grass species as herbage for livestock in Canterbury. It had involved field trials in part of what is now Hagley Park in central Christchurch. The experiment came to an abrupt halt when the plots were ploughed under.[34] A well-known Canterbury farmer and frequent commentator on farming matters in the daily press, Marmaduke Dixon, criticised the methods used by farmers to establish pasture on their properties and advocated an alternative system that 'will preserve a great many of the native grasses, and your paddock all the year round will be a good stay and a resort for the stock, markedly in the winter time, when the artificial pastures ought to be almost closed'.[35] In its October 1913 issue, the *New Zealand Farmer, Stock and Station Journal* printed a short report on a lecture by Mr Wakeman to the Pahiatua Agricultural and Pastoral Association in which he described his success in feeding finely cut toetoe to his livestock, and outlined how this native plant could be cultivated. He also reported excellent results from feeding young leaves of cabbage tree to beef cattle.

Debate about the respective merits of native and introduced grasses continued throughout the second half of the nineteenth and into the early twentieth centuries, and in 1879, J. B. Armstrong published a brief account of ten native species: 'not one of the ten but is equal to the average of the English grasses usually cultivated for the purposes of permanent pasture'.[36] Armstrong mentioned the group of experts appointed by the Canterbury Philosophical Institute 'to inquire into the merits of the various native grasses', and ruefully conceded that 'the labours of the committee did not induce farmers and others interested to pay any attention to the matter' owing, he believed, to the difficulties many individuals had experienced while trying to identify any of the recommended native grasses that might be found growing on their properties. He anticipated that John Buchanan's manual,[37] with its beautiful, detailed illustrations of native grasses, would resolve this problem. Armstrong was undeniably partisan in his endorsement of native grasses on farm and station, but the following statement shows his knowledge of them: 'I am rather of the opinion that the colony possesses very few native grasses suitable for short lays, but for permanent pastures we have many which will eventually prove equal to the best of the imported ones.' He called for a carefully controlled, experimental approach to the selection, propagation and management of palatable and nutritious native grasses on New Zealand farms. In 1901, James Wilson reviewed Buchanan's manual and discussed the merits of three of the most promising native species:

Danthonia pilosa (= *Rytidosperma racemosum*, now thought to be a naturalised Australian species[38]), *Danthonia semiannularis* (= *Rytidosperma setifolium*) and *Microlaena stipoides*. During the unusually dry summer of 1897 these three had reportedly spread and thrived in places where introduced grasses had failed, and were recognised as palatable to sheep and cattle as well as drought resistant. Their principal drawback, which Wilson acknowledged, was that their seeds had proven difficult to harvest in commercially viable quantities and were expensive compared with those of perennial ryegrass and the other pasture plants of commerce.[39]

PROSPECTS FOR NATIVE SPECIES IN THE HUMANISED LANDSCAPES OF SOUTHERN NEW ZEALAND

Almost from the start of organised settlement, there had been warnings about the magnitude and consequences of forest clearance, but none for the removal of the tussock or the draining of wetlands, and they appeared in local as well as overseas newspapers, magazines and books. Nothing in the farm and station diaries I read, however, suggested that these warnings had been noted, let alone acted upon, by many members of the first generation of European settlers. Those who did respond tended to be the well-established holders of large pastoral properties like the Acland, Potts and Tripp families rather than the occupants of smaller farms. The view remained strong in settler society that native species were unlikely to persist when in competition with introduced plants and animals. It was a convenient opinion, for it meant that farm and station holders did not have to justify their wholesale transformation of the vegetation cover. Even J. B. Armstrong lent support to this view, and in 1871, barely two decades after organised settlement began, he added the following words to his list of 180 naturalised plants in Canterbury: 'The indigenous Flora seems to have arrived at a period of its existence, when it has no longer strength to maintain its own against the invading races [of plants]; indeed, every person who has attempted the cultivation of native plants knows how difficult it is to cultivate most of them, on account of their weakness of constitution.'[40] This may be the origin of a widespread misunderstanding about native New Zealand plants that not even the strenuous advocacy of Leonard Cockayne could dispel,[41] and it condemned them to playing the role of museum exhibits:

relicts of a fast-disappearing world that might be seen growing in reserves and botanical gardens but not in significant numbers in the humanised landscapes of southern New Zealand.

A change in attitude was, however, becoming evident in the final three decades of the nineteenth century, when many farm and station diaries contained references to visits by family members to forest remnants to collect ferns and decorative shrubs for planting around the house.[42] By the close of the nineteenth century, many rural properties in southern New Zealand had only scattered remnants of the old vegetation cover: typically, tussock on higher slopes, shrubland in steep-sided valleys and small patches of wetland on the flats. Canon James Stack, appalled by what he saw at Goughs Bay and elsewhere on Banks Peninsula, published this epitaph for a world lost in his own lifetime:

> Barbarous vandalism and a desperate greed for every blade of grass has spoiled the beauty of many a [Banks] Peninsula home, and efforts that are now being made to raise plantations of pinus insignis [*Pinus radiata*] and other trees show what a wise thing it would have been to have spared a few patches of that unrivalled native bush that, once destroyed, no art can replace.[43]

Four decades earlier, Stack had been struck by the park-like landscapes of Inch Clutha, south Otago, noting the 'native forest trees, which the good taste of the settlers had preserved, the beauty of which was very much enhanced by the cultivation of the ground around them'.[44] Despite the strenuous advocacy of influential people like William Pember Reeves, it took almost a century for the national psyche to come to terms with the fact that, if given the opportunity, many native species can hold their own, thrive and improve the utility, value and appearance of farms, stations and settlements in the humanised landscapes of southern New Zealand. In 1913 a handbook described ways to transplant native species from bush to garden, and in 1975 an article was published in the widely read *New Zealand Journal of Agriculture* extolling the benefits of remnant native bush on pastoral properties 'as shelter for stock against cold winds and driving rain'.[45]

CHAPTER FIVE

In with the New
Introduced Plants and Grazing Animals

The following account of introduced economic plant and animal species in the lowlands and low hill country of southern New Zealand is rooted in the emerging ideas and practices of the agricultural, biological and environmental sciences. By the 1850s agronomy, animal husbandry, horticulture, pasture management, and plant and animal selection either were or were becoming established scientific disciplines in Great Britain, Western Europe and the United States, and Charles Darwin's books on evolution were widely available. As will be discussed in this chapter, the rudiments of ecological, environmental and evolutionary thinking were known as early as the 1860s in New Zealand, and they influenced the nature of environmental transformation in the south.

Scarcely 40 years after the start of organised settlement in the province, James Stack described the lowland landscapes of north Canterbury as 'studded in all directions with comfortable homesteads surrounded by cornfields and well-stocked pastures' and noted the speed with which this had happened: 'So rapid has been the process of transformation, that persons who have come to these shores within the last twenty-five years have found everything about them so like what they left behind in the Old World that the change of residence has proved to them more like a removal from one English county to another than removal to a foreign land.'[1]

European settlers transformed the lowland and low hill country landscapes of southern New Zealand with pastures for sheep and cattle, mown lawns, vegetable gardens and plots of decorative plants, orchards and berry gardens, hedges and shelter belts, and trees and shrubs for wood and decoration. The outcome was artifice in league with nature. All were systems of mostly introduced plant and animal species in which the normal ecological functions had been attenuated and the forces of natural selection suppressed by propagation of preferred varieties, cultivation and weeding, pruning and replanting, control of insect pests and diseases, mulching and fertilising. Some contained more and others fewer species than the evolutionarily tested ensembles of plants and animals that they replaced, but all were akin to early successional ecological systems and required close maintenance if they were to persist.

Insofar as grazing lands for sheep and cattle were concerned, European settlers believed that the long-term carrying capacity of southern New Zealand's indigenous grassy and shrubby ecosystems was too low for economic viability, so they strove to replace them with judicious selections from the many commercially available, fast-growing and palatable imported grasses and broadleaf herbs. There was another consideration: in some places the tussock grasses were so tall that sheep and cattle could be overtopped. A flock of sheep can disappear from view in a tract of snowgrass, which would have raised serious problems for recent settlers whose land lacked fences or, in areas like the Canterbury Plains, high points from which they could look down on their properties. All a landholder might see when looking for his livestock was an undifferentiated stretch of tussock grassland, and it is probable that this featured in the decision to set fire to vast tracts of the Canterbury Plains in the early 1850s. It was not just about spurring the perennial grasses to produce soft new shoots after their tops were destroyed. A settler's overriding objective was to establish as speedily as was practicable an artificial system that would facilitate animal management, produce nutritious herbage for as long as possible during the frost-free season, support sheep and cattle throughout the year, and, wherever it could be achieved, ensure a crop of hay during late spring or early summer.

SETTLERS AND ADVISORS THINKING ECOLOGICALLY

In Western Europe, the United States of America and the British colonial territories, ecological ideas were implicit during the 1870s and 1880s and beginning to be widely known by the end of the century. The terms 'dominant', 'succession' and 'diversity' are core concepts of plant ecology and were evident in the expert advice given to New Zealand landholders during the late nineteenth and early twentieth centuries. Early in the twentieth century the scientific publications of Alfred Cockayne, who became Director General of Agriculture, and his successor Bruce Levy formalised and extended expert ecological advice to farmers. Cockayne was an early advocate of applying successional theory to pasture development: 'These changes that occur in pastures are known as successions, and their study is probably the most interesting of any phase of pasture investigation. A complete knowledge of the factors affecting succession would have very far-reaching effects on our management of grass-land, but up to the present they are in many cases quite unknown.'[2] Bruce Levy went further and, in a conference presentation on seed mixtures for managed pastures, he opened with a brief account of successional development in native New Zealand forest and tussock grasslands before extending his ecological argument to sown pastures.[3] His four conclusions drew on his extensive experience as a professional agronomist and, although formally expressed in the new terminology of plant ecology, they were akin to what farmers had been advised since the late nineteenth century: first, 'the more closely plant demand and environment harmonize, the simpler the association becomes and the simpler can the seed mixture be that is employed'; secondly, 'in sowing down to permanent pasture, phases in the succession [leading] to an ultimate dominant should be provided for'; thirdly, 'for the ploughable country where one can govern the habitat and management at will, the case is strong for a simple seed-mixture composed of one or two grasses and one or two clovers . . . . It is of great practical significance that the grasses and the clovers are ecologically related, and no [pasture seed] mixture-maker can avoid this fact'; and, fourthly, 'control of the environment is presumed before simplification of seed-mixtures can be carried out'.

Plant and animal species fill distinctive niches in an ecological system, where they occupy a stream of long-wave radiation and have access to water, nutrients and room to grow. The landscapes of farm, station and rural district are patchwork quilts of interdependent, mostly artificial, ecological systems.

One ecological model views a landscape as a mosaic of nested environmental patches at different stages of ecological development after a disruptive event such as landslide, fire, flood, disease or death of a plant from old age. Provided a patch of environmentally disturbed ground has been neither directly nor indirectly affected by people, a relatively small number of fast-growing but short-lived plant species will occupy it and facilitate the establishment of other species. In time, a mature ensemble of longer-lived plants will develop. Regardless of whether the landholder sowed grasses and clovers on a ploughed field, scattered them between the shoot crowns of burned tussocks or allowed spontaneous regeneration, the outcome would be a system likely to experience spontaneous change in vegetation composition and structure.

Which species become established in a large or small area of environmental disturbance depends on what survived on-site or grew nearby. In a region of intensive agriculture dominated by periodically ploughed fields, local reserves of seeds and other propagules will be steadily depleted and some species could even disappear from the regional landscape. As settlers found, however, this may not isolate cultivated land from those native and introduced weedy plants that persist on roadsides and in patches of environmentally disturbed land on their properties, ready to spread from there to ploughed fields and pastures. Weedy species are mostly early pioneer plants and their preferred niches are areas of disturbed ground where they may compete with crop and pasture plants for nutrients, water and room to grow. In many respects, an expanse of cultivated ground is a battlefield occupied by early successional species: some are highly valued by landholders, and others are nuisances; some are palatable to livestock, and others not; some depend upon people to sow, nurture and harvest them, and others are able to complete their lifecycles and persist for several generations without human agency.

There are few analogues in nature for an agricultural landscape comprising the extensive environmentally homogeneous, early successional patches that we know as cultivated fields. Little wonder, therefore, that from the early years of organised European settlement in southern New Zealand farmers and station holders battled to control infestations of weedy plants and pest animals. Darwinian principles barely apply to the managed ecological systems of sown pastures and crops on ploughed fields, leaving them without an evolutionary future, and the principal activities of agriculture and animal husbandry cut across the important ecosystem qualities of self-regulation, nutrient and water cycling, and balance between resource inputs and outputs. A further consequence, as some commentators in settler society recognised, is that whenever wool, milk, meat, hides or harvested plant tissues are sent off the property, an environmental cost is incurred by the area's reserves of energy, nutrients and water.

PASTURE ESTABLISHMENT AND MANAGEMENT

Oats were grown almost everywhere draught horses were used on pioneer properties for cultivation, ploughing and harvesting; and fields of wheat were common on lowland properties during the 1860s, 1870s and 1880s because of wheat's value in providing the food staple flour. Joseph Davidson, who operated a mixed cropping and livestock farm on the outskirts of Waikaia in northern Southland, grew about twenty acres of wheat for sale each year to a nearby flour mill, and other settlers did the same. From the earliest years of organised settlement, rural people grew sufficient potatoes to satisfy their own demands and to feed the pigs kept for slaughtering in winter. Some, like the residents of The Point Station in mid-Canterbury, sold excess production to neighbouring runholders or transported sacks of potatoes by wagon to large and small towns for sale to residents. Settlers with large flocks of sheep and numerous cattle had difficulty making sufficient hay for supplementary winter feeding, so wherever they could be grown turnips and swedes were planted for consumption by ewes in the weeks before lambing. Turnip, swede and rape seed were often added to mixtures of grass and clover seeds for sowing in cultivated land intended for new pasture. The most important short-lived introduced plants were, however, grasses and clovers.

Almost from the start of organised settlement it was evident to settlers that grassland farmers enjoyed two significant advantages over their peers in Great Britain: sown pastures could be grazed almost year round, and only relatively small amounts of supplementary feed had to be set aside for the normal winter.[4] In 1913, Alfred Cockayne highlighted the advantages enjoyed by pastoral farmers in New Zealand when he noted that 30 per cent of British pastures were cut for hay, and that during the growing season 40 per cent of all cultivated land was reserved for hay and other animal feed.[5] By 1861 there were 158,000 acres of sown pasture in New Zealand, but twenty years later that area had reached

3.5 million acres, so rapid was the transformation from native to managed systems of plants and animals.[6] Amongst the reasons for this transformation was the fact that in the lowlands, a sown pasture could support three sheep to the acre year round compared with one sheep to an acre of prime tussock grassland.

In transforming the vegetation cover of their properties, it took several decades for settlers to come to terms with the conjunction of climate, small-scale topography and soil type as a guide to the types of pasture best suited to the large- and small-area environments of their properties. In 1902 a locally published farming journal reprinted with evident approval the advice of an eminent British agronomist, Professor John Wrightson, about seed mixtures for English farms: 'The mixtures should be of such a nature as to suit the land and the climate, for climate is not a mere matter of latitude, but also of longitude and altitude. Climate varies, even on a farm, to an extent which affects the period for sowing and reaping. Local climate often changes within a mile, because it is affected by aspect and shelter; and soils vary from field to field, or even from acre to acre of the same field.'[7]

The documentary record of settlers' transformation of the open landscapes of the lowlands and foothills of southern New Zealand into enclosed landscapes of productive pastures for sheep and cattle is one of swirling tensions and ongoing debates. There were seven topics that drew particular discussion: first, the relative merits of introduced and native edible plants; secondly, the value of local experience compared with knowledge acquired elsewhere; thirdly, the comparative advantage of complex over simple seed mixtures; fourthly, whether it is better to sow large or small amounts of pasture plant seeds; fifthly, whether a short duration pasture is a more realistic goal than a permanent pasture; sixthly, the benefits of cheap over expensive seed mixtures; and finally, the desirability of including one or more dominant species in a mixture of productive pasture plants. Of those topics, the importance of high quality of seed was the most significant. Unlike grassland farmers in the United States of America, pastoral farmers in New Zealand were frequently informed that clean, ripe seeds might be expensive but that they repaid the additional cost after a few years. In its December 1904 issue, the *New Zealand Farmer* reprinted a long article from the *Cape Colony Journal of Agriculture* on the importance of sowing good quality seed whenever a pasture was established. With the aid of a worked example, the author illustrated how 'cheap seed often costs more than nominally dearer but intrinsically better seed'. The causes of difficulty for farmers and station holders alike were pollutants such as weed seeds, dust and chaff as well as the often large numbers of dead seeds of the desired species. The solution involved varietal selection and rigorous control over pollutants, supported by germination trials conducted by the farmer on each batch of seed before it was sown. Buyers were particularly advised against placing undue faith in a vendor's assurances of quality.

Although usually termed 'English grasses' – during the nineteenth century, those two words were applied to palatable broadleafed herbs as well as to pasture grasses – the introduced species selected for New Zealand pastures were of diverse geographical provenance. Ryegrass grew wild in Great Britain before it was adopted then adapted as a grass for sown pastures. Clover appears to have entered European pastoral farming from the Middle East and North Africa via Andalusia, having been cultivated there by Moorish farmers.[8] In words that reveal his strong ecological thinking, Bruce Levy wrote that 'each type of pasture decidedly has a life-history – a life-history subject to modification under the many vicissitudes of soil, climate, and environment, made complex or simplified according to the species concerned in the pasture association, and by the method of utilization of the feed produced'.[9] In the continuation of that essay, Levy made much of the fact that 'steep hillsides afford many more types of succession. Here the factors at work are more potent, and the need for greater knowledge more pressing . . . . An undesirable succession will mean reduced carrying capacity, and in many cases the complete abandonment of those areas for a long period of years, until such time, in fact, as a fresh covering of natural vegetation, such as fern, &c, has been produced, enabling the area to be fired and a fresh seed-bed thus secured.'[10]

Permanent or Short-duration Pastures?

In 1898, a Mr P. Patullo was struck by the failed attempts to establish permanent pastures in Canterbury, and commented critically on the large area then sown in poor quality introduced grasses that had reverted to inferior swards before the cost of pasture establishment had been recouped. He concluded that 'experiments in the direction of proving the most suitable grasses and the management of them after they were sown would surely be worth attention'. As for the degraded pastures of Crown-owned pastoral runs in Central Otago, he argued that 'it would pay government to experiment with various grasses, either native or imported, to see if a fresh growth could be induced to come'.[11]

There were clear indications in nineteenth- and early twentieth-century newspapers and farming magazines that New Zealand farmers were anxious to achieve a long-lived pasture yet unwilling or unable to allow sufficient time for it to mature before opening it up for intensive grazing. A precept of the day was that it takes a farmer a lifetime to achieve a true permanent pasture that will steadily improve with age and not be invaded by such undesirable species as Yorkshire fog. One commentator decried what he termed the disgraceful state of sown pastures in New Zealand, and argued that even in Canterbury a permanent pasture could be achieved by sowing sufficient seed of several commercially available varieties, including ample clover, and grazing the developing pasture judiciously: 'The great object in sowing a variety [of grass and clover seeds] is that they spring to maturity at different times, the stock will always have a good bite (weather permitting), whereas by sowing only one sort, say rye grass, which is very commonly done, there is no succession of food.'[12] Andrew Simson reported that in England it was not unusual for a 'sole' – 'mature turf' is a fair approximation of that term – of grass and broadleafed herbs to take twenty years to form. He castigated settlers for their poor preparation of seed beds in immature agricultural soils, use of inferior seeds and excessive reliance on ryegrass. A consequence of these lapses was that inferior pastures had to be ploughed up and resown more frequently in New Zealand than in England.[13] A little later Sir James Wilson reminded his readers about the slow pace of development of grassy turfs in nature and urged farmers not to be in too great a hurry.[14]

Dominant Species and the Need for Nutrients

On 22 February 1868 the *Otago Witness* reprinted an article that had first appeared in the *Canterbury Times* under the headline 'Growing Grass for Sheep'. It noted that a mixture of perennial ryegrass and clover was almost invariably sown in the province, that the usefulness of prairie grass was compromised by the poor quality of seed then commercially available, and that 'what we want here for pasturing in summer is a grass that will send down its roots deep into the subsoil so that it will not be affected by the often long-continued droughts and hot winds. The cocksfoot grass seems to answer all the requirement of the case . . . . Sow more cocksfoot and less ryegrass, and don't be niggardly with the above seed, and don't overstock for the first year.' That advice was not uniformly accepted, and on 6 September 1887 the manager of Ida Valley Station wrote to Nimmo and Bell, seed merchants in Dunedin: 'I wish to lay down grass, seven and a half acres, which eventually will be chiefly used for making into hay. Please send me sufficient seeds (mixed) of the kinds your experience has taught you is best for this purpose and a note of the kinds (with prices) for future reference. Put no cocksfoot in. Send also enough rye [grass] and tares mixed to sow an acre as I want to try it – also five pounds of lucerne.'[15] His willingness to experiment with a range of pasture plants and be guided by a commercial organisation are noteworthy.

In his order placed with the Canterbury Farmers' Co-operative Association in Timaru, and billed by that company on 30 September 1913, James Preston, then based at Centrewood Station in east Otago, showed his understanding of the desirability of sowing two complementary dominant grasses (in this case, perennial ryegrass and Italian ryegrass), one fine grass (crested dog's-tail), a legume (cowgrass, a perennial form of red clover) and a root crop (swedes) to tide sheep over during the first year, and of spreading a balanced mineral fertiliser with the seed to ensure strong plant growth during pasture establishment. He paid £21 1s for four tons of superphosphate and bone dust, and £70 17s for fourteen sacks of ryegrass, seventeen sacks of Italian ryegrass, 300 pounds of cowgrass, 200 pounds of crested dog's-tail and 56 pounds of swede seeds.[16] Phosphorus and nitrogen are key mineral nutrients for healthy plant growth, and the sulphur in superphosphate is commonly needed in the brown and grey soils of hill and mountain country.

Perennial Ryegrass as a Pasture Dominant

As early as 1877, Robert Wilkin acknowledged the considerable soil nutrient demands of perennial ryegrass and its tendency to become ergotised but came to the conclusion 'that no single grass will fatten stock so quickly as rye grass'.[17] He believed that a portion of it should be added to the seed mixture for every sown pasture. To balance that advice, however, he concluded his survey of common pasture plants in Canterbury with the opinion 'that no single grass will ever make a good pasture'. In 1889, 'Ovis' commented on the desirability of perennial ryegrass in sown pastures and referred to the controversies then current about its role in pastoral farming. He argued that it must be the leading, if not the only, grass in temporary pastures, and declared that 'ryegrass in its best form is a most valuable grass, but there are different varieties of it, some of which are inferior. For reasons which I need not enter into here, variety [species richness] is indispensible, and the greater that variety the better.'[18] During the 1890s pioneer farmers' experience in the recently cleared bush country of the North Island showed that perennial ryegrass did not persist in a sown pasture and was accordingly better treated as a semi-permanent grass.[19] In 1890, Cockayne had identified the four great advantages of perennial ryegrass as a major component of sown pastures in New Zealand: cheap seed, ready germination under most environmental conditions, high yielding on most arable land and well liked by livestock. Its prime disadvantages, however, were that it required moist fertile soils to persist, its flowering stems were rejected by livestock when plant tissues they preferred were available, and new shoot growth was slow between the peak of flowering and start of the next growing season.[20]

In the nineteenth and early twentieth centuries, the 'perennial ryegrass'[21] available from seed merchants was found to have serious shortcomings.[22] As Cockayne and others had discovered, very little of the commercially available seed was from long-lived perennial plants. It also suffered from indifferent mechanical cleaning and tended to come with undesirable weedy plant pollutants. Despite their acknowledged palatability and nutritional value, even acknowledged perennial varieties of ryegrass needed well-watered, fertile soils if they were to persist. When mature plants died, they left small patches of bare ground for other plants, especially weedy adventives, to occupy, and in environmentally challenging situations perennial ryegrass did not tend to regenerate spontaneously. These undesirable attributes attracted considerable attention from pasture scientists, amongst them Levy who argued that shoot die-back was usually the consequence of poor pasture management rather than some intrinsic feature of perennial ryegrass. Settlers were made aware of these shortcomings and strongly advised to sow only the best quality seeds from a reputable supplier, never to buy already mixed seeds but rather to order them separately bagged for mixing on the property, to seek written assurances from the merchant about germination rates and to conduct simple germination trials on damp sacking before sowing the seeds.

I found in their diaries and personal papers that settlers occasionally asked for 'ryegrass seed harvested from old pastures' when placing an order with a merchant. Agents for imported and locally grown seeds usually made much of where they had been harvested and whether or not they came from old plants, the latter indicating the strong likelihood of a true perennial rather than a shorter-lived variety. A British variety, 'Devon Evergreen', was known as perennial and may have been introduced to Poverty Bay by missionaries associated with the Reverend Samuel Marsden. In the nineteenth century, several companies in southern New Zealand advertised 'Poverty Bay ryegrass', and the seeds commanded premium prices. Advertisements published in the *Otago Witness* between 1851 and 1900 referred to 29 named varieties and geographical provenances of ryegrass seeds, a number greatly in excess of that for any other sown pasture plant at the time and an indication of the importance of this matter to farmers. An early critic of perennial ryegrass as a dominant species in sown pastures in New Zealand warned farmers against sowing it on light soils or in colder areas where it might suffer from frosts and cold winds. He also reported that timothy thrived in wet as well as dry ground and generally did better than perennial ryegrass.[23] The following decade 'Ovis' noted that in Canterbury perennial ryegrass did best on moist fertile soils where winters were not particularly cold and drought was unusual. He also advocated that it should be sown in admixture with the seeds of a deeper rooting pasture plant like cocksfoot and a clover.[24]

In the 1890s an influential British farmer and agricultural commentator, Faunce De Laune, described perennial ryegrass in notably disparaging terms, and his judgement was re-published throughout New Zealand. In its 6 February 1885 edition, the *Akaroa Mail* reprinted an extensive piece taken from the *Rangitikei Advocate* in which De Laune's remarks were quoted verbatim. The unsolicited publicity gained by producers in two local centres of cocksfoot seed production was presumably appreciated. Sir James Wilson of Bulls, a frequent contributor to the *New Zealand Farmer*, had found on his Rangitikei farm that meadow foxtail and timothy were superior to perennial ryegrass on prime soils and to cocksfoot on second-class land. Alfred Cockayne contributed to the debate when he wrote, 'My personal opinion is that the value of ryegrass has been exaggerated on all classes of country except those of the very highest quality.'[25] Later in that essay he attributed the historical popularity of perennial ryegrass amongst New Zealand farmers to the cheapness of its seed.

The 'ryegrass question' engaged several influential local commentators but I was unable to find documentary evidence to suggest that farmers in southern New Zealand had bought into the argument. If anything, they showed their allegiance to a well-known pasture plant despite the ebb and flow of a very public debate. An allied feature, and one that certainly did influence local farmers, was the steadily growing recognition that local practice may sometimes be worth more than advice from overseas, as the following example illustrates.

Although pastoral farmers in the United States of America accorded high status to Kentucky blue grass, their peers in New Zealand recognised it as a weed that was almost impossible to eradicate from a sown pasture once it had become established.[26] As 'Ovis' warned, 'Mr Sutton clearly recognises the fact that a grass which may be valuable in one country may become a pest under different conditions of soil and climate. He gives as an indication the case of *Poa pratensis*, which is considered a useful grass in Britain, and one of the most valuable of grasses in parts of America, but which in New Zealand assumes the character of a troublesome twitch.'[27]

By the close of the nineteenth century, farmers, station holders and their advisors had come to recognise the necessity of local field trials before adopting a new pasture plant, even one that came with strong endorsement from farmers overseas, because New Zealand was already showing its potential to harbour populations of introduced pasture plants in natural and disturbed habitats where many were about to become serious environmental weeds. One contender to replace perennial ryegrass as a pasture dominant was meadow grass, a favourite of Cockayne, but its seeds were relatively expensive and difficult to obtain in bulk, and timothy had its supporters. A little later, Primrose McConnell concluded that 'Prairie grass (*Bromus uniolides*) may probably in the near future take the place of rye-grass in permanent pastures, being of a more permanent nature.'[28] But it was cocksfoot that almost carried the day.

Like all introduced pasture plants, cocksfoot had its strengths and weaknesses. It was relatively slow to establish after sowing, and its tussock growth habit required careful grazing management to ensure the availability of palatable herbage. Those disadvantageous features were evident to a correspondent from the Hawke's Bay, who wrote about the management regime needed to keep the coarse growth of cocksfoot under control on that area's medium-quality soils, where it did well.[29] Despite its disadvantages, however, cocksfoot was a palatable and nutritious perennial species that remained productive over dry periods in summer, a consequence of its deep rooting system that was able to tap subsurface water and nutrients.

Diverse or Simple Seed Mixtures?

Another matter that called for attention related to the number and identity of palatable and nutritious species in the seed mixture for a sown pasture. Early in the twentieth century, New Zealand pastoral farmers sowed about 40 species of grass and ten of clover – many more than their counterparts in North America, South Africa and Australia – although far fewer were sown on a single property.[30] In New Zealand the debate swung back and forth during the second half of the nineteenth century and in the years leading to the First World War, as it had in Great Britain since the mid-eighteenth century. A widely held view amongst settlers and their advisors was that a diverse ensemble of edible species provided a measure of insurance against drought, disease or old age affecting one or more species in the sown mixture. In its 23 March 1867 issue, the *Weekly Press* reviewed the place of cocksfoot in sown pastures and came down firmly on the side of mixtures of several pasture species to ensure grazing for livestock and to improve soil quality. In England during the eighteenth century there was broad support for two-species mixtures of pasture plants – typically perennial ryegrass and either white or red clover – because they ensured abundant, nutritious and palatable herbage on well-watered, fertile soils during the frost-free season. The contrary opinion held that perennial ryegrass tended to grow relatively slowly during the height of summer and was demanding of water and nutrients in the topsoil. The two-species combination came to be seen as flawed because it did not offer sufficient variety to grazing animals. That led to a dominant pasture plant being the favoured model. A preferred alternative was a mix of several less frequent species with complementary growth surges to ensure variety throughout the growing season, but it proved almost impossible to achieve: admirable in theory, but challenging in practice.

Entries in nineteenth-century farm diaries and letter books showed the extent to which grassland farmers heeded the call for diverse seed mixtures. On 1 October 1868 one of the Teschemaker brothers noted in his diary that he had sown a mixture of seven pasture plant species in former tussock country on his south Canterbury property: cocksfoot, Italian ryegrass, meadow foxtail, timothy, alsike, and white and yellow clover.[31] He opened his diary for 1873 with a note in which he recorded that he had recently sown 500 acres with a slightly different mixture of seven species, at an application rate of 45 pounds per acre. Six years later, on 23 September 1879, John Finlay who farmed at Makikihi in south Canterbury received a consignment of seeds for sowing in newly drained and ploughed land: twenty bushels of ryegrass, four pounds each of red and white clover, and four pounds each of cowgrass and 'Marybel' wurzels.[32] At that time, some farmers were sowing several species of clover to ensure that at least one of them would survive an environmentally difficult season. The virtues of a clover were its ready establishment, rapid growth rate, high nutritional values and palatability, tolerance of grazing and trampling, and enhancement of soil nitrogen content through the activity of nitrogen-fixing bacteria in its root nodules.

Some palatable herb species were included in small quantities in seed mixtures to provide 'bite' for sheep and cattle – plants like chicory, plantain, vetch and yarrow – until experience showed that several of them occupied considerable areas in a sown pasture and produced relatively little herbage in return. The last three decades of the nineteenth century were times of continuous evaluation of seed mixtures for artificial pastures, and most issues of magazines intended for farmers included recipes for an extraordinary range of seed mixtures, each especially formulated for the small or large area environments of a particular property.

Early in the twentieth century, the *New Zealand Country Journal* was advising its readers: 'Grass mixtures cannot be standardized; varieties that suit and prove thoroughly satisfactory on some soils and in some districts, prove very much the reverse when tried in other districts, or even in other classes of soil in the same district.'[33] The article extolled local experience and urged each farmer to experiment with different seed mixtures before deciding which species to sow in a ploughed field. This was evident in the order that James Preston placed in 1901 for another of his properties, Kyeburn Station in Central Otago: one mixture of seeds was for 430 acres of pasture and the other for 160 acres.

TABLE 5.1 James Preston's order for grass seeds, 1901.

	First mixture	Second mixture
Ryegrass	35 lb / acre	30 lb / acre
Cocksfoot	2	2
Italian ryegrass	–	4
Chewing's fescue	2	–
Crested dog's-tail	1	1
White clover	2	2
Timothy	–	2

SOURCE: INFORMATION EXTRACTED FROM JAMES PRESTON'S PAPERS; SEE APPENDIX NOTE 4

In 1918, Cockayne reported the advantages of mixed over pure sown pastures: they offered a variety of feed to livestock, yielded better over the growing season, were less affected by short-duration unfavourable conditions for plant growth, and allowed valued pasture species to be incorporated where they would not do as well if grown alone.[34] A later commentator, J. M. Smith, distanced himself from Professor Stapleton's opinion that on poor soils a farmer could do little better than sow only white clover; and on average quality or better land, where white clover might reasonably be expected to volunteer, sow only ryegrass. To Smith, however, 'a mixed pasture provides a measure of insurance against abnormal conditions that cannot be provided for in pure or nearly pure sowings'.[35] Simple seed mixtures or complex, the debate was not confined to New Zealand. It was a topic for long-standing discussion in British agronomical circles, and was glossed in a reference book published during the 1930s.[36] Experience in New Zealand, however, was already showing that while complex seed mixtures might persist in moist fertile soils, on poorer ground or in areas prone to drought in summer, a single grass species – and it was usually not the species most palatable to sheep and cattle – eventually dominated the sward.[37]

Weight of Seed Sown

An early lesson for pioneer grassland farmers was that for a sown pasture to be successful, great care is required when preparing the seed bed: ploughing to the right depth in winter, clearing away plant scraps, allowing frost to break up the clods of topsoil, rolling to consolidate the seed bed and evenly sowing the requisite weight of thoroughly mixed high-quality seeds. White clover was found to be a desirable component of the seed mixture for sown pastures, with the prime exception of pastures intended to last only a year or two in summer-dry parts of the country. For the latter, either red or subterranean clover was preferred, and expert advice was that their seeds had to be sown at a rate of three or more pounds per acre. The recommended weight of sown seeds increased steadily during the late nineteenth and early twentieth centuries, and peaked during the First World War. During the late 1890s the average weight of grass and clover seeds sown was about twenty pounds per acre, but by 1916 the national average was closer to 30 pounds per acre.[38]

Variety and Provenance

In 1952, Elizabeth Gray published a short historical account of perennial ryegrass in Scottish pastures.[39] The documentary record began in the late seventeenth century, and perennial ryegrass has been widely planted since then.

First woolshed at Holme Station, south Canterbury, late nineteenth century, with the homestead at the rear and the hired men's quarters in the middle distance. SOUTH CANTERBURY MUSEUM, TIMARU, 1899

As one British authority rather guardedly put it, perennial ryegrass was 'not altogether desirable, but not of no value'. The original seed seems to have come from wild populations in England, and the Scottish county of Ayrshire became known as a source of high-quality seed for the local and export trades. During the 1860s about 3500 tons of ryegrass seed, which was recognised as a perennial form, were harvested annually. Many cultivated varieties were known to British farmers, amongst them 'Antonshill', 'Molle', 'Pacey', 'Pollexfen', 'Roughead', 'Russell', 'Stickney', 'Stirling', 'Sutton' and 'Whitworth' ryegrass. All were in demand, and several were advertised in New Zealand during the second half of the nineteenth century.

Although large quantities of pasture plant seeds were imported, the emerging preference was for locally grown material, one of the earliest of which was white clover seed harvested on New Zealand properties. Imported seed might have lost viability during the long voyage by sailing ship out to New Zealand followed by the further delay until sowing, and pastoral farmers began to favour seeds sourced from parts of the colony known for the excellence of their product. High-quality cocksfoot seeds, for example, were produced by growers on Banks Peninsula ('Akaroa' cocksfoot) and at Sanson in the Manawatu, but farmers were also advised to harvest seeds from the most successful pasture plants on their own properties and to sow the seeds when re-establishing a pasture. This was an expression of Darwinian thinking, and it pervaded the advice given to farmers during the last quarter of the nineteenth century. As Professor Hutton mused in print, 'Farmers are in the habit of obtaining seed from other localities, as it is found that by changing the conditions under which plants are grown, greater vigour is obtained. But Mr Darwin suggests that it would be still better if the farmer obtained only half the seed necessary and mixed it with an equal quantity of the same variety grown on his own farm. This would secure cross-fertilisation and the yield would be larger.'[40]

The advent of varietal selection, seed certification, improved methods of soil cultivation, top-dressing with mineral fertilisers, irrigation and better grazing management enabled agriculturalists to smooth over any small-area environmental variations on their properties, leading gradually and eventually to the application of ecological theories to pasture establishment and management in colonial New Zealand.

INTRODUCED PLANTS FOR BOUNDARY MARKERS AND SHELTER

Few farmers in New Zealand have any idea of the time, the labour, and the expense required to produce a good hedge. In my part of Scotland, in an elevated position, it requires about nine years to rear a thorn or beech hedge into a secure fence . . . . In lower and more favourable situations, I suppose four or five years would be required. After the fence is established, it ought, besides being pruned, to be cleared of weeds about the roots at least once a year. This is a point never attended to in New Zealand, and consequently the lower branches get rotten, and the hedge becomes open at the bottom.[41]

The lowland and low hill country landscapes of southern New Zealand where settlers established farms and stations were mostly open and grassy, with shrubs in cooler and moister sites, cabbage trees in damp depressions, and patches of forest in gullies as well as on the higher slopes.[42] As Charlotte Godley – wife of Robert Godley, the first administrator of the Canterbury Settlement – recognised, without ready access to native trees a settler family would struggle to build house and out-buildings, cook food and heat water, let alone install fences to control stock movements.[43] Absence of wood on a property was acutely felt in places where there was neither stone for construction nor peat for household fires, and in its 11 April 1868 issue the *Otago Witness* reported the cautionary words of Mr Murison, a member of the Tokomairiro Farmers' Club in Milton:

> I am confident that time will show that the fact of our [Otago] agricultural districts having originally possessed merely a sufficient quantity of bush to supply the wants of the first settlers is a circumstance which will eventually prove detrimental to the interests of our farmers unless steps are taken by them to carry on sylviculture upon a more extensive system than they are now doing.

Fence posts, rails, strainers, hurdles and pit-sawn lumber could be brought in from distant stands of forest, but it was to a landholder's advantage to establish groves of fast-growing woody plants to satisfy the family's needs for fuel, construction and fencing materials. Until those trees matured, residents were dependent on what they could obtain locally or buy from suppliers. Among the important environmental services provided by plantings of trees and tall shrubs were shelter for humans, livestock and crops, as well as decoration of homesteads. The likelihood of a pay-off was usually good within three or four years of planting, although the range of benefits might not become evident for a decade.

Insofar as hedges were concerned, they were usually formed from gorse, broom and hawthorn, with boxthorn more common in the north of the South Island. In 1877, 'Agricola' published notes about five possible hedge plants for Canterbury – bramble, gorse, hakea, hawthorn and kangaroo acacia – describing hakea as the 'best hedge plant ever tried in this province'.[44] Given the aggressive weedy tendencies of several species of this Australian genus in southern Africa, later generations can only be grateful that it was not widely adopted by settlers in Canterbury. Several native New Zealand species were known to be well-suited to hedging,[45] but there is scant documentary evidence of their use. Landholders recognised the benefits and took risks when establishing hedges for boundary marking, livestock shelter on windy sites and grazing control. In its 27 September 1862 issue, the *Otago Witness* reprinted with evident approval a short piece on this topic from an Australian newspaper, the *Yeoman*: 'The cultivation of live fences is a matter of great importance. Post-and-rail fences are chiefly objectionable as affording no shelter either to crops or stock, and shelter, as we have often observed, is a question of very great importance in these [Australian] colonies.'

Even while they were painstakingly sowing gorse seeds along surveyed fence lines or on sod walls, and tending to young plants, many landholders in the lowlands of Canterbury and Otago were grubbing out adventive gorse plants from fields where hedges had been established several years earlier. So bad had the threat posed by adventive gorse plants become by the 1880s that on 20 April 1889 the Clutha County Council formally resolved to petition its sitting member of the House of Representatives 'to get a clause inserted in the Local Government Act to prohibit gorse from being sown alongside country roads [in the county] where it does not already exist'.[46] Amongst the other problems, a gorse hedge that was allowed to sprawl over fields and road verges created a congenial habitat for rabbits in areas of farmland.

In well-wooded tracts like eastern Southland, landholders could afford to divide their properties into fields surrounded by post-and-rail fences, although most also installed hedges for shelter, but where there was sufficient but not an abundance of wood in the locality, landholders usually erected post-and-five-wire fences. In the second half of the nineteenth century, thick iron wire was commonly used for fencing but was bulky to transport, difficult to install and expensive. When eight-gauge galvanised wire and light barbed wire become readily available, secure fences could be quickly installed, and at a reasonable price, around and through the property. A further strategy was to establish post-and-wire fences on 50 to 75 centimetre-high banks of piled up topsoil or sod cut from the turf on either side of the projected fence line before sowing gorse and broom seeds or planting rooted hawthorn cuttings – the latter were referred to as 'quicks' in plant merchants' catalogues – on top of the bank. Sod banks were useful short-term fences, but when gorse and broom were planted on them they also became attractive habitats for rabbits. As the hedge plants grew, leader shoots were woven through the wires and side shoots were trimmed to ensure a barrier sufficiently tall and dense to deter sheep and cattle from breaking through. When the hedge was well established, fence posts and wire could be removed for recycling elsewhere on the property.

A live hedge required regular clipping to maintain it in good estate, and its flowering and fruiting shoots had to be stripped to reduce the likelihood of seed dispersal. The ledger books for Te Waimate Station in south Canterbury contained numerous entries relating to gorse hedges on the property.[47] The earliest full year in the extant volumes was 1878, when labourers working under contract cut and stripped 67 kilometres of gorse hedge and cleared gorse seedlings from seventeen kilometres of road verge. Over the next ten years, contract labour was hired annually to trim and strip between three and 29 kilometres of gorse hedge, and to clear gorse seedlings from between one and ten kilometres of road verge. The ledger entries did not indicate if men employed as general labourers on the station also cut hedges or grubbed seedlings.

The first two generations of European settlers on the eastern flanks of the South Island were assiduous in planting large numbers of commercially available herbaceous and woody plants on their properties. On 4 April 1868 the writer of a letter to *The Press* predicted that in a few years 'the exposed Canterbury Plains...will be shaded with walnut and other English forest trees'. On Otaio Station in south Canterbury, the Teschemakers planted 5406 young trees in September 1879: 2800 bluegums, 1626 macrocarpa and 980 radiata pine.[48] The outcome of tree planting across southern New Zealand was a progressively enclosed countryside dominated by large numbers of introduced tree and shrub species, with few native plants and animals in the mix, making rural properties places of relatively high biotic diversity, albeit of mostly exotic species: in fields, around houses, and alongside tracks and roads.[49] The situation changed as rural people discovered which of the many introduced species best supported the farm economy and which were unnecessary to, or prejudiced, the operation. This, along with the widespread use of mineral fertilisers and senescence in trees and shrubs planted a century earlier, triggered an overall decline in biodiversity on most rural properties during the second half of the twentieth century.

Every large town had at least one merchant who sold seeds, cuttings and rooted plants in his own right or on commission, and these well-connected tradesmen imported material from Australia, Pacific Rim countries, Britain, Western Europe and farther afield. The defining features of their imports were variety and numbers, and in those regards pioneer New Zealand was not alone in

Married shepherd's cottage, Raincliff Station, south Canterbury, late nineteenth century.
SOUTH CANTERBURY MUSEUM, TIMARU, 1985

the colonised world of the late nineteenth century. A recently published article in *National Geographic* compared numbers of recognised horticultural varieties of food plants commercially available to home gardeners in the United States of America in 1903 with those on sale 80 years later: for example, from 544 named varieties of cabbage in 1903 to just 28 in 1983, from 497 to 27 named varieties of lettuce, and from 808 to 79 named varieties of tomato.[50] Advertisements placed in local newspapers by merchants in late Victorian New Zealand listed the many varieties of the crop, food and decorative plants then commercially available, and much the same was evident in the documentary records of farm and station families. On 17 July 1865, John Wither 'planted 18 forest trees between the house and the water closet' (on Otematata Station), and one week later he 'planted a quantity of willows'. Another settler in the open country of southern

TABLE 5.2 Plants bought from merchants and established on Longlands Station in the Maniototo by James Preston between 1892 and 1899.

Trees and shrubs for shelter and decoration
Balsam poplar [*Populus* sp.]
Bluegum [*Eucalyptus* spp.]
Cupressus macrocarpa
Lawsoniana [*Chamaecyparis lawsoniana*]
Poplar [*Populus* spp.]
Oak [*Quercus* spp.]
Larch [*Larix* spp.]
P. douglasii [*Pseudotsuga menziesii*]
P. larix [unknown]
Pine [*Pinus* spp.]
Pinus maritima [*Pinus pinaster*]
Torreyana [*Pinus torreyana*]
Pinus insignis [*Pinus radiata*]
Sequoia sempervirens
Spruce [*Picea* spp.]
Thuja [Arborvitae, *Thuja* spp.]
Weeping willow [*Salix babylonica*]
Wellingtonia [*Sequoiadendron giganteum*]
Willow [*Salix* spp.]
+ unspecified 'trees'
Hedge plants
Briar [*Rubus* spp.]
Broom [*Cytisus scoparius*]
Gorse [*Ulex europaeus*]
Garden and orchard plants
Blackcurrant [*Ribes nigrum*]
Gooseberry [*Ribes* spp.]
Ivy [*Hedera* spp.]
Laurel [*Laurus nobilis*]
Plum [*Prunus* spp.]
Rhubarb [*Rheum* sp.]
Veronica [*Veronica* spp.]

SOURCE: INFORMATION EXTRACTED FROM JAMES PRESTON'S PAPERS; SEE APPENDIX NOTE 4

New Zealand, James Preston, brought in numerous shelter, hedge, decorative and food plants (Table 5.2). Where they are known, or can be reasonably inferred, formal botanical names are in italics.

The large numbers of individual plants involved are also evident in an order placed on 12 April 1883 by the manager of Ida Valley Station in Central Otago with Gordon Brothers, owners of Braidwood Nursery in Dunedin, for 6314 shelter and decorative tree seedlings and saplings (Table 5.3).

TABLE 5.3 A single order of trees and shrubs for planting at Ida Valley Station, Central Otago in 1883, arranged in decreasing order of numbers of plants.

Number of plants	Name of tree or shrub
2200	*Cupressus macrocarpa*
1500	Norway spruce [*Picea abies*]
1050	Pinus insignis [*Pinus radiata*]
500	*Pinus pinaster*
200	Cupressus lawsoniana [*Chamaecyparis lawsoniana*]
200	Douglas spruce (sic) [*Pseudotsuga menziesii*]
200	Laurel [*Laurus nobilis*]
100	Betula [possibly *Betula pedula*]
100	Larix europea [*Larix decidua*]
100	Oak [*Quercus* sp]
50	Wellingtonia [*Sequoiadendron giganteum*]
24	Holly [probably *Ilex aquifolium*]
24	Laurestinia [possibly *Viburnum tinus*]
24	Mountain ash [either *Eucalyptus regnans* or *Sorbus aucuparia*]
12	*Cupressus funebris*
12	Thuja gigantea [probably *Thuja plicata*]
12	Thujopsis borealis [possibly *Thujopsis dolabrata*]
6	Weeping [willow] [*Salix babylonica*]

SOURCE: INFORMATION EXTRACTED FROM THE MANAGER'S LETTER BOOKS: SEE APPENDIX NOTE 8

PLANTS FOR THE HOMESTEAD

The administrator of the Canterbury Settlement, Henry Sewell, and a Christchurch resident from 1853 to 1856, characterised the Canterbury Plains as open, dusty, windy, exposed, stormy and bleak.[51] Those and other words were used by settlers to describe their new homes in the years before they could assemble more congenial living areas. Rural and townspeople alike progressively transformed the landscape and environments of their properties with mown lawns, vegetable and flower gardens, imported grasses and palatable herbs, orchards, hedges, plantations of decorative and useful trees, and thickets of shrubs. Plantings around homestead and out-buildings provided shelter from strong winds and winter cold and, just as importantly, ensured a measure of psychological security for family members and employees alike: a human-scaled, artificial environment in the midst of what must have seemed at the time a vast open expanse stretching without break between distant horizons. A thick green wall of macrocarpa trees was established like a hollow square around many country cottages to keep out livestock and create a snug retreat for residents.

An example of this transformation is in The Point journal from a pastoral property on the high plains and low hill country of mid-Canterbury for the period between August 1867 and April 1871.[52] Although it had been first settled by Europeans in the early 1850s, there was little more than a rough wooden shelter and a few kilometres of post-and-wire fence in 1862 when the Phillips family from nearby Rockwood Station took over the lease. Two of the Phillips boys, a cadet and several hired men moved in during August 1867 and immediately began to transform the property. An early imperative was to improve the living quarters, install a sheep dip and shearing shed, and develop a pig sty and storage sheds. As that work progressed, they began to fence off the homestead from wandering livestock, sow a lawn, plant vegetable and flower gardens within the protected area, and establish windbreaks as well as an orchard. Their diaries contain the names of plants recently established at The Point, and the only native species that the writers identified were brushwood – probably mānuka twigs for staking peas and beans, and to provide protection to young garden plants from the wind – and assorted but unnamed plants from a nearby area of native forest that were taken for use as specimen plants, firewood, fencing materials and sawn wood.

Other farm diaries for the period also recorded trips by residents to nearby stands of bush to collect soft and tree ferns, flowering shrubs such as hebe and cabbage tree, as well as curious shrubs and flowering vines such as the native pohue (*Clematis* spp.) for transplanting to a garden on the sheltered side of the house. In the early 1850s one early New Zealand resident, Constantine Dillon, had a small property on the Waimea Plains near Nelson and, writing to family in England, mentioned native New Zealand plants and occasionally included seeds of clematis, flax and kākā beak. He also sent a collection of named dried ferns to his mother.[53]

Analyses of diaries and letter books from ten rural properties in north Otago and lowland Canterbury, each dating from the first three decades of European settlement, showed that out of a total of 180 purposefully or accidentally introduced non-native plant species, 41 per cent were plantation and specimen trees, 22 per cent were fruits and vegetables, 16 per cent were crop and pasture plants, and 11 per cent were decorative plants for flower gardens and shrubberies. Of those remaining, 6 per cent were aggressive weeds such as Californian, Scotch and sow thistle, couch grass, groundsel, sheep sorrel and vetch, and 4 per cent were hedge plants such as broom and gorse.[54] In his diaries Constantine Dillon listed 23 types of vegetable and pot herb that he had planted in an acre of vegetable garden during the 1850s.[55] Contemporary newspaper advertisements listed the many fruit trees then commercially available – several varieties of apricot, cherry, mulberry, nectarine, peach, pear and quince. The diaries of The Point Station did not include names of the horticultural varieties and cultivars brought onto the property, but referred to training apple trees in espalier form along sheltered, north-facing walls, which the author of a newspaper article published in the early 1850s had described as the best way to ensure that fruit trees would withstand the area's strong winds.[56] In 1884 the manager of Ida Valley Station sent the following order for fruit trees to E. Fulton of Dunedin: four each of desert apples and cherries, two each of cooking apples and red plums, and one each of greengage, egg plum, desert pear and cooking pear. Amongst the favourite nut trees at that time were filbert and walnut. Newspaper advertisements also indicated the great variety of flowering plants then available to the interested country resident.

In 1905, James Miller, who had a farm in the Outram–Maungatua district of east Otago, listed in his diaries the varieties of perennial plant growing in his berry garden and orchard.[57] Long as it is, the list is representative of the times (Table 5.4, overleaf).

TABLE 5.4 Numbers of horticultural varieties of tree and berry fruit grown on James Miller's farm near Outram in east Otago. A question mark signifies an unknown number of plants.

Number of plants	Horticultural variety
50	'Laxton's Noble' strawberries
26	'Carter's Prolific' raspberry canes
24	redcurrant bushes
16	gooseberry bushes
12	whitecurrant bushes
9	'Kentish Hero' blackcurrant bushes
8	'Carter's Black Champion' blackcurrant bushes
8	'Lee's Prolific' blackcurrant bushes
8	'Pretty Boy' gooseberry bushes
1	Japanese plum tree
1	quince tree
?	more than 25 varieties of apple
?	7 varieties of pear
?	6 varieties of peach
?	6 varieties of plum trees
?	4 varieties of apricot

SOURCE: INFORMATION EXTRACTED FROM JAMES MILLER'S DIARIES; SEE NOTE 57

INTRODUCED GRAZING ANIMALS

Most of the farm and station diaries available to me were kept during the second half of the nineteenth century. Aside from references to merino, Leicester and crossbreed, few mentioned sheep or cattle breeds. Early in the period of organised settlement, merino were the basis of some of the first flocks, especially in the drier hill country and inland basins. There was some interest in matching flock type with prevailing environmental conditions, and as early as 12 June 1867 the *Bruce Herald* drew on an article first published in the *North British Agriculturalist* that reported a speech by Mr J. J. Mechi to members of the Faringdon Farmers Club: 'One of the most important causes of success in agriculture is to adapt your crops and animals to the climate and soils. An infringement of this natural law is sure to be attended with loss.'

Dairy cows, Waimate district, south Canterbury, showing original post-and-rail and more recent post-and-wire fences, and factory-built gates: early twentieth century.
2001-1026-00065, WAIMATE HISTORICAL SOCIETY AND MUSEUM

Settlers discovered that no one breed of sheep or cattle was well suited to the range of environmental conditions found nationally, regionally or even on a single large property,[58] but for more than three decades the national flock was predominately merino, Leicester, Lincoln and crossbreeds.[59] Crossbreeds became more popular as pastoral farmers shifted to producing both meat and wool for distant markets,[60] and a long period of experimentation began. By 1900, however, several once common sheep breeds had virtually disappeared from New Zealand farms.

During the West Coast and Otago goldrushes, sheep and cattle were driven to the burgeoning mining settlements to satisfy the demand for meat. In his diaries, Edward Chudleigh referred to driving cattle into the Otago goldfields inland from Palmerston as well as across Lewis Pass to the West Coast,[61] and residents of The Point Station in mid-Canterbury drove herds of cattle across Arthur's Pass to the West Coast for sale to miners in the Hokitika goldfields.[62] These were two of many such instances. Settlers who lived closer to towns supplied fat livestock to the local butcher. One was Michael Studholme of Te Waimate Station in south Canterbury, and records of his monthly sales for the years 1882–88, inclusive, have been preserved.[63] This period straddles the close of the wool era and the start of the frozen meat and dairy produce era,

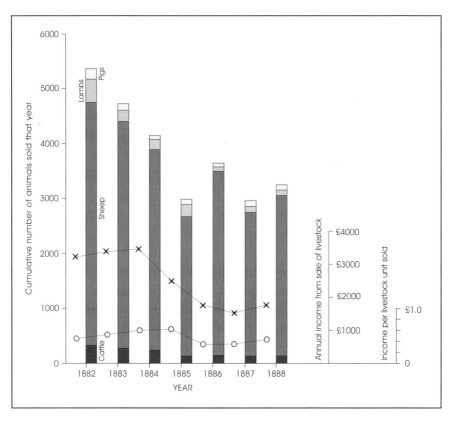

FIGURE 5.1 Cattle, sheep, lambs and pigs sold by Michael Studholme of Te Waimate Station to a butcher in Waimate, south Canterbury, between 1882 and 1888. The lines connecting hollow circles show changes in average annual value per livestock unit of these sales, and the lines connecting crosses show changes in annual income from animal sales over the period. SOURCE: INFORMATION EXTRACTED FROM AN INCOMPLETE RUN OF LEDGER BOOKS MAINTAINED BY MANCHESTER AND GOLDSMITH; SEE NOTE 47

when the country was shifting from its earlier economic dependence on exports of fine and crossbreed wool to a diversified suite of primary products – many of which required refrigerated shipping – for northern hemisphere markets. Ledger entries gave monthly numbers of sheep, lambs, pigs and cattle sold to a Waimate butcher, and the payment received. I used the following values to convert numbers of animals into livestock units: 1 for sheep and lambs, 3 for pigs and 7 for cattle, and those values are plotted in Figure 5.1.

In September 1885, William Shirres wrote to his financial backers in London about the poor return for exports of frozen meat. He noted that it was selling on the London market for between 4d and 4½d per pound, although it had recently reached 5d, whereas local sales of meat were fetching between 2½d and 2¾d a pound. Freight from Oamaru to London (he did not mention the cost of slaughtering and freezing) for a carcass was 2½d per pound, 'pointing to the miniscule profit from exporting'.[64]

The principal cattle breeds were Hereford and shorthorn, and pastoral farmers were reluctant to adopt other breeds until later in the nineteenth century.[65] The situation changed, however, as the export market for dairy products opened up. Wilson documented importations of Jersey, Guernsey and Friesian cattle, all of which resulted in increased production of milk solids on farms with dairy cows, and owners of dairy herds benefited from contacts with their neighbours: 'Early generations of dairy breeders . . . . saw the commercial results of their matings in the [milk] bucket twice a day. They discussed with all the other producers carting milk or cream to the local dairy factory the reasons why some had more cans on their drays than others.'[66]

CONCLUSION

Within a decade of occupying their properties, pioneers in seasonally dry areas like the Canterbury Plains realised that sown pastures would deteriorate within two or three years regardless of the care they put into establishing them. This was not necessarily the case elsewhere in southern New Zealand: in western Southland, for instance, multi-species, sown pastures twenty years of age and older were known. One strategy employed by nineteenth-century farmers to achieve a 'permanent pasture' involved attempts to mimic nature by sowing mixtures of pasture species from a range of life forms and with complementary growth rhythms in the expectation that the mature emergent grasses would provide shelter for a sub-canopy of smaller grasses, while a ground layer of palatable broadleafed herbs and fine grasses would cover the topsoil and guard against the establishment of undesirable species. Experience, however, led Bruce Levy to declare, 'Theoretically, the ideal pasture would be that which contained a number of species whose seasonal growth varied so that there would be provided a uniform and continuous growth throughout the whole year. In practice such a pasture is exceedingly difficult to secure.'[67] The problem, as he showed, was that of the 40 or so species of pasture grasses and clovers commonly planted in New Zealand, most put on their greatest growth in spring and early summer, and only between five and sixteen of them, depending on the place, showed even moderate growth during winter.

The difficulties of identifying plants with varying growth seasons were compounded by ecological factors. A native ecosystem comprises several species of plants and animals; typically has multi-layered pyramids of numbers and biomass; and is characterised by a complex food web comprising green plants, herbivores, omnivores and carnivores, as well as innumerable microorganisms that occupy the soil and litter layers where they break down dead plant and animal tissues and facilitate nutrient cycling. In sown pastures, domestic sheep and cattle eat a broad range of tissues from several species, but these are not notable features of a native ecosystem. The combinations of palatable plants and grazing animals established on pioneer farms and stations lacked those co-evolved traits that can ensure the persistence and resilience of an ecological system. Furthermore, by cultivating, fertilising and draining their land, settlers lessened the effects of, and even suppressed, fine-grained environmental variations that had been occupied by distinctive ensembles of plants and animals before their properties were environmentally transformed.

Settlers showed interest in decorative plants native to New Zealand, and several were widely planted around their homes, yet despite there being suitable species in the indigenous flora, settlers depended on introduced shrubby plants for hedges, and introduced trees for fuel, fence posts, lumber and shelter. They showed little or no interest in harvesting indigenous plant foods. Seeds, bulbs, rooted cuttings and saplings were imported and propagated, and from the 1880s onwards settlers were advised to assure themselves of the quality of newly purchased seeds by conducting simple germination trials before sowing them. Later, measures to ensure quality control through seed inspection and certification of pasture plants were instituted and field trials were conducted to provide quality assurance for farm and station.

Establishment of humanised landscapes in southern New Zealand involved replacement of native by introduced species and development of alien ecosystems. Persistent tussock grassland communities were supplanted by short-lived soft grasses and clovers to allow an economy of rapid growth in livestock. Landholders then sent as much as possible of that biomass off the property to national and international markets. Native ecosystems characterised by slow turnover in biomass and large reserves of plant growth nutrients and water were replaced by managed ecosystems with rapid turnover in biomass and relatively small reserves of nutrients and water. In some areas, these reserves proved too small and had to be supplemented by inputs of nutrients in the form of fertilisers and by irrigation water to ensure an acceptable economic return to the property holder. Landholders aspired to a landscape that would work to a different rhythm: a decade or less for a full cycle compared with many decades in the tussock grassland and low shrubby ecosystems that once covered the land. Speed was the operative word: prompt germination, rapid early establishment, fast growth of palatable tissues, short generation times for pasture plants and the animals that grazed them, and regular production of seeds and offspring for another generation. Perennial ryegrass, cocksfoot, timothy, meadow grasses and fescues, as well as red and white clover had been bred to occupy this niche and were well suited to the role. When introduced pasture plants succeeded, farms and stations prospered, but success came at a cost to settlers, and some of those costs were not recognised for several years. Livestock compacted the surface soil, and where they overgrazed parts of the pasture or broke through the sod with their hooves, they opened the area to invasion by undesirable weedy plants and rabbits. Unless they were painstakingly managed, artificial pastures had to be replaced after three to ten years, and the combination of trampling and frequent cultivation destroyed the soil profile. A rapidly cycling pastoral system is akin to a finely tuned engine and requires continuous maintenance to keep it running as designed, protect it from unusual environmental conditions and guard against break down. It also needs high-quality inputs to keep it running. Few settlers achieved all these goals.

CHAPTER SIX

Emerging Environmental Problems Erosion and Declining Soil Fertility, Pest Animals and Weedy Plants

In the late 1870s, S. Grant and J. S. Foster, visiting New Zealand on behalf of fellow tenant farmers in Lincolnshire, were struck by this country's diverse soils, the variety of which, they reported, 'is so great that a detailed description is impossible; adjacent farms in almost every district may vary almost completely'.[1] They did not comment on erosion, but noted that poorly yielding fields 'turned out on inquiry to have been cropped with wheat without any manure for the last four or five years',[2] and were struck by 'turnips, grass, thistles and docks all growing together with almost equal profusion' on farms in drier parts of north Otago.[3] Rabbits were not widespread at the time of their visit, although they reported that banks alongside the railway line leading into Invercargill were honeycombed with burrows.[4] Elsewhere in Southland they found that 'rabbits are a serious nuisance.... There is some talk of exterminating them with phosphorous, but whether the plan be feasible or not we cannot say.'[5]

If they were to prosper, settlers needed to respond appropriately to the environmental variety, potential for erosion and burgeoning problems with weedy plants and pest animals such as those observed by the two visiting English farmers. Grant and Foster were able to see with the eyes of outsiders some of what was happening to the land. For the settlers there was not the distance or the immediate ability to make comparisons. Instead, they required good powers of observation, the will and resources to respond promptly and effectively to each new environmental challenge, the capacity to deal flexibly with difficulties, the knowledge and skills needed for innovation, and commitment to the task of establishing an economically viable holding.

Before a reliable communications infrastructure was in operation, amongst the greatest challenges faced by rural settlers were those that stemmed from the capacity of the New Zealand environment to throw up surprise after surprise. For 50 years, the most frequently addressed topics in their diaries were the state of the day's weather and its consequences for farm or station, but from the 1870s onwards, rabbits and the methods settlers employed to eradicate them were described in progressively greater detail. Weedy plants became an issue with the transformation of the tussock grasslands and were significant in the 1880s and later. Environmental and economic issues relating to soil stability and fertility crept up on rural people, and were seldom noted in their diaries during the first 25 years of organised settlement. What intending settlers had heard or read before embarking from Great Britain left them ill-prepared for problems posed by soils of mostly moderate fertility, unstable geology, and introduced plant and animal pests. If organised settlement had begun 50 years later, then settlers might have found greater guidance and clearer warnings in the emerging principles of ecology and environmental science. Instead, they assembled novel ecosystems before they could have known about the predictions of these new sciences. It was a prodigious experiment – a term that settlers knew and used – conducted without controls or hypotheses, and few participants appreciated the long-term risks to their respective properties once it was under way.

Introduced plants and animals dominated the thinking and practice of successive generations of European landholders as they strove to develop economically viable operations in the low shrub and tussock grasslands of southern New Zealand. For a decade, native species had been sources of food for settlers and livestock, as well as wood for construction, fencing and fuel. Large and small stands of native bush on the property or nearby provided lumber and fuel as well as shelter and browse for cattle; tracts of tussock grassland fed and sheltered ewes at lambing time, especially when a heavy snowfall made it difficult for owners to muster their flocks to lower ground; and wetlands were sources of reeds and rushes with which settlers could thatch farm buildings or stacks of hay, oats and wheat held in the barnyard until they could be processed. Many introduced species were purposefully brought onto a property by settlers, and some became naturalised there. Others arrived as pollutants in sacks of seeds or lime,[6] as well as on the pelts of livestock bought at stock sales. And a relatively small number spread spontaneously from their places of establishment, naturalised in the diverse environments of a large rural property, formed viable populations and were recognised as weeds when their potential for economic disruption overreached any perceived value they might have had at the time of introduction. Amongst the latter were broom, hawthorn and gorse, ferrets, rabbits and sparrows. Several herbaceous weeds, such as sheep sorrel and three species of thistle, spread rapidly and established dense populations in the environmentally disrupted habitats of modified tussock grasslands and cultivated land, where they soon affected economic viability.

WORSENING EROSION AND DECLINING SOIL FERTILITY

From the earliest days of European settlement pioneers had been bothered by blown sediment and had frequently suffered debilitating eye infections as a consequence, but there is no evidence in their diaries for a significantly greater incidence of dust storms during the period of tussock clearance. If anything, the problem appears to have eased when introduced grasses and clovers took root in ploughed or burned land and established a close sward over the topsoil. One early commentator argued that European settlers 'felt that they had got possession of a grateful country' and that 'The basis of the remarkable prosperity is undoubtedly to be found in the fertility of the soil and the delightful climate of New Zealand',[7] but entries in settlers' diaries frequently tell a different story. An early reference to accelerated erosion in a tract of gully woodland on Banks Peninsula comes from Joseph Munnings' diary entry for 27 April 1860:

> Blew a gale from S.W. all last night and today, accompanied with torrents of rain without intermission all day. Water came through [the roof] and onto my bed and every room in the house. A great deal of damage was done about S.W. end of Gov[ernor's] Bay . . . . About half a mile up Mr P's gully a slip of earth came down

(some 30 or 40 tons of earth, stones, etc.) off the side of the gully and stopped the water which accumulated, and then it slipped a little further and soon it got to the beach and carried everything before it; excepting the largest trees, all the shrubs and small trees quite cleared out.[8]

Two years later, on 24 May 1862, the *Otago Witness* reported that heavy rain over several days in Dunedin had caused slumping at building sites: 'If the rain continues we shall not be surprised to hear of some serious accidents by land slips.' On 13 November 1864, John Wither noted in his diary: 'A showery day. Was out the length of the paddock [on Otematata Station]. A great deal of damage to the fences [after the heavy rain and flooding]. Thousands of cart[load]s of earth and stones carried down the gullies.'

The 17 February 1868 edition of the *West Coast Times* reported the start of what was to be a year of appalling weather and serious erosion across southern New Zealand: 'There has been a landslip of some 400 or 500 tons of earth about a quarter of a mile above Carey's House at Blueskin . . . . Cherry Farm at Waikouaiti has been completely inundated, and at least 1000 sheep drowned. There have been numerous landslips along the road . . . . The Taieri Plain is reported like a sea many miles wide.' On 23 July 1868, Wither recorded 'Heavy showers during the night and morning . . . . A good deal of the [creek] bank slipped in.' After he had relocated his family to a property on the flanks of Lake Wakatipu from a hill country station several kilometres north of Arrowtown, Wither's reports of flood damage grew more frequent, as this sequence of daily entries from 25 to 28 September 1878 illustrates:

A wet night and heavy rain all day. The creeks all very high; the highest flood I have seen.
A dry forenoon; very wet in the afternoon. The creek like a river, and sweeping a great deal of soil away, and shifting its course.
A dry day nearly all day. Rain at night. The creeks very high, and a grand quantity of ground washed away. Slips off the hills in all directions . . . . The bridge off the Lockie, and water running all over.
A showery day, and a high flood set in in the afternoon.

Back in its 9 August 1868 edition, the *Bruce Herald* had reprinted the following news item from the *Lake Wakatip Mail*: 'We are informed the number of land slips and earth falls in the Shotover District has been very remarkable during the past few months.' The correspondent associated these events with an unusually high frequency of earthquakes, and they had similar consequences across southern New Zealand. Wither's station lay nearby, and on 29 August there was 'A tremendous flood this morning. The ground all on the move. A fearful night of rain and lightning. The creeks tearing away banks and fences and shifting its banks. Shifted all perishables on to the upper floor of the shed [on the shore of Lake Wakatipu]. The Lake up to the doorstep of the shed.' At the end of the year, the 5 December issue of the *Otago Witness* reported a landslip at Nevis Bluff on the Queenstown Road where 'Some thousands of tons of rock fell down'.

Severe flooding and erosion affected southern New Zealand again a decade later, and on 12 October 1878 Wither recorded, 'A showery day . . . . The creeks very high and washing away the banks in several places.' Damage to his property was widespread, and twelve days later he wrote how he 'Went round by Halfway Bay [on Lake Wakatipu]. The river has cut away the banks.' His next references to flooding were in the second half of the 1880s, following several years of below-average rainfall. By then, he was attempting to mitigate the adverse effects of episodic flooding on his property by carting large stones into a creek that had been eroding its banks. The floods of 1888 caused serious disruption for Wither, notably on 31 March: 'A very high flood. The highest for 2 years. Water in all directions. The big creek washing away its banks and changing its course.' Four days later he described 'A blowy day; heavy rain at night. Creek [channel] turned, stones raised [to deepen the channel], rushes cut [to widen the channel]; a great deal of water running off the hills.' Depletion of the tussock cover in the uplands had evidently reduced the amount of water that could be held in short-term storage, resulting in rapid runoff into streams and rivers after heavy rain and thawing of the winter snowpack. That winter, between 25 and 29 August, he described 'placing bush lawyer vines and scrub [possibly matagouri] into the creek to hold stones for bank protection', which he repeated on 1 September, 24 November, several times in December and again on 22 November 1889. The 1890s were marked by periodically severe flooding and erosion across southern New Zealand, and on 1 December 1891, Joseph Davidson declared 'the ground [was] too wet after the heavy rain last night . . . . the [Waikaia] River and [Dome] Creek were too dirty [for fishing] so they got nothing.'

Erosion and nutrient depletion go hand-in-hand, with vegetation clearance the normal precursor to soil loss. It is not the case, however, that early settlers were unaware of the need to replenish the soil. Even the earliest residents of farms and stations manured their potato crops and vegetable gardens with material gathered from mucking out barns and pig sties and clearing rotting plant tissues from the bases of haystacks, something that Māori cultivators found strange.[9] But within a short time landholders had begun to realise that few New Zealand soils were especially fertile.

On 3 April 1863, *The Press* reported that a shipload of guano had recently been unloaded in Southland, and two years later, on 8 March 1865, it reprinted a long article from the *Australasian* about the utility of phosphatic fertilisers to farmers. On 23 October 1865, Edward Chudleigh recorded that he had 'planted potatoes [at Mount Peel Station] in the afternoon, putting a lamb's tail to every potato. They ought to grow well.'[10] Early public warnings about declining soil fertility were published in the *Bruce Herald*, and on 13 November 1867 it printed a reader's letter about likely soil impoverishment resulting from the actions of unaware farmers. The writer endorsed the editor's earlier call for a halt to burning-off after the grain harvest. Ten days later the same newspaper printed a detailed summary of a lecture on the topic 'Exhaustion of Soils' that had been delivered by an elected member of the Otago Provincial Council to the Tokomairiro Farmers' Club. The topics of soil conservation and soil fertility figured frequently in that newspaper's pages, and on 4 November 1868 it published a letter from 'Enquirer' who asked 'Can you inform me if bone dust is good to use for growing turnips, and what quantity is proper to apply per acre, and further if it would be an advantage to mix guano with it.' In reply, the editor advised three alternative applications: either six hundredweight (approximately 300 kilograms) of bone dust, or three hundredweight of bone dust mixed with manure, or a mixture of four hundredweight of bone dust and two hundredweight of guano. In this period, wood ash, guano, bone dust and ground lime were spread on ploughed land and around root crops, but widespread use of mineral fertilisers did not eventuate until the 1880s.

There were differing opinions about the fertility of New Zealand's soils, akin to the suppositions about the weather and the experience of settlers on the land. A contrary opinion had appeared in a handbook reviewed by the *Otago Witness* in its issue of 26 September 1868. The publication, directed to British readers, was the work of the Otago Provincial Council – *The Province of Otago, in New Zealand: Its Progress, Present Conditions and Prospects* – and in it much was made of the 'tropical fertility of the soil'. Twelve years later an independent assessment was published by S. Grant and J. S. Foster who had toured rural New Zealand in the late 1870s. They reported with some surprise that 'No artificial manure is used there [in the wheat fields in Canterbury] or, as far as we could gather, elsewhere in New Zealand.'[11] An even more extreme opinion was expressed by the author of the section on agriculture, published in the *New Zealand Official Yearbook* for 1894: 'So full is the soil of plant food that several continuous crops of potatoes and cereals may be taken with little apparent exhaustion [of the nutrient supply].'[12]

Awareness of the importance of maintaining soil fertility is nonetheless evident in landholders' diary entries. On 12 August 1876, John Wither recorded that 'Fissey [had been] digging bones round the fruit trees'. The following decade, on 24 September 1887, he mentioned that he had 'Sowed bone dust in garden', and two weeks later he noted that he had 'Wheeled [wood] ashes into garden', a traditional way to maintain adequate levels of potassium in garden soils. Deficiency of this macro-nutrient may explain why Joseph Davidson recorded in his diary on 12 May 1890: 'Commenced to take up the potatoes. They are a very poor crop – very small.'[13]

In a remarkably prescient article, Robert Wilkin added his voice to the debate about whether land deteriorated or improved under continuous sheep grazing. He illustrated his argument by reference to a theoretical example – a farm from which 1000 sheep and 5000 pounds of wool are sent away each year – and posed the interrogative question 'what is removed from the farm land through these sales, and where does it come from?'

> [It] must, in the first instance have come from the soil, and as a large proportion of what is taken away by sheep consists of mineral matter, it could scarcely be derived from the atmosphere. The disintegration of the rock by the effect of the atmosphere and by the continual trituration of the sheep's feet may supply a portion of the waste, but I cannot imagine that anything like the full quantity is supplied in this way, and therefore trust that the question may be taken up by men who are capable of telling the readers of the JOURNAL (*sic*) what is actually taken away, and how the waste is supplied.[14]

By the final decade of the nineteenth century or the start of the twentieth, environmental and economic problems raised by low and declining soil fertility had become strikingly evident, and farmers could buy bulk supplies of ground lime and guano as well as a wide range of imported mineral fertilisers. On 6 March 1905 the manager of the Gore branch of New Zealand Loan & Mercantile Agency reported to the manager of the Invercargill office that he was holding stocks of the following mineral and organic fertilisers: Triumph No. 1 turnip manure, Triumph No. 1 grain manure, Malden Island guano, Lawe's superphosphate, Kainit (potash imported from Germany for root crops), Kempthorne and Prosser manure, and Southland Frozen Meat Company manure.[15] In 1930 Herbert Guthrie-Smith summarised several decades of observations on his hill country station, Tutira, in Hawke's Bay where, by 1908, 'the countryside was intrinsically worth less for grazing purposes than it had been in 1879, when grasses were first sown'. He concluded that the signs of diminished soil fertility were clear in the historically recent reduction in proportions of ryegrass and white clover plants, the decreased carrying capacity of sown pastures and the shallow roots of sown English grasses.[16] In his later publications about New Zealand pastures, the agronomist Bruce Levy often highlighted the situation:

> It has been the custom for farmers to abuse the inferior grasses and clovers that come into their pastures and which, according to them, 'oust the better grasses.' The plain truth, I think, is this: that the better grasses and clovers have been starved through lack of plant-food, and so weakened thereby that the entry of inferior grasses and low-fertility-demanding weeds was made possible. If the farmer will only feed his better grasses and clovers sufficiently with artificial manures &c., there is virtually no danger of their ever being replaced by inferior grasses and weeds.[17]

RABBITS

The first release of the European rabbit appears to have been near Riverton in Southland during the late 1840s, but on two occasions during January 1858 William Macdonald of Orari Station in south Canterbury released rabbits on the island between the two branches of the Rangitata River.[18] Across southern New Zealand, however, the principal releases were during the late 1850s and 1860s.[19] Within twenty years of liberation, rabbits had moved inland, except where their passage was halted by large rivers, and were progressively occupying the more open dry tussock grasslands and low shrub country of northern

Southland, Central Otago and the Mackenzie Country.[20] As early as 12 July 1863, Edward Chudleigh could record 'All hands gone to Rangitata Island to shoot rabbits', almost certainly for recreation and the pot.[21] According to W. McLean, who was employed for many years as a rabbit exterminator in the Wairarapa, rabbits will survive on most types of pasture but require 'bare ground on which to rear their young'.[22] Despite the risk of drowning during floods, riverbanks are good habitats for rabbits, but he observed that they will burrow freely in open areas with friable subsoil. On average, six adult rabbits were thought to eat the same amount of herbage as one sheep.[23]

The Rabbit Nuisance Act 1881 set up rabbit control districts – eight in the North Island and fourteen in the South – each with residential inspectors who regularly visited affected properties to observe and report on the effectiveness of control measures. In 1886 the Joint Committee on the Rabbit and Sheep Acts heard about the recent spread of rabbits into the provincial district of Canterbury from the Mackenzie Country in the south and west, and the Amuri district in the north, as well as the comparably rapid range expansion into the King Country and Hawke's Bay.[24] That same year the Joint Rabbit Nuisance Committee heard evidence, including the words of one resident of Waimea South: 'I do not think the [sale of rabbit] skins pay[s] us 35 per cent of the cost of destruction.'[25]

In 1887 interest was expressed in controlling rabbit numbers by a contagious disease, and one suggestion was to approach Louis Pasteur for expert advice.[26] The so-called 'swine plague' was lethal to rabbits in Canada, but concern was felt in New Zealand that it and 'red measles' could affect domestic pigs and, possibly, sheep. The following year a fuller report was released on methods of rabbit control, and it named parasitic diseases found during autopsies of dead rabbits: a bladder worm, a liver parasite, rabbit mange and lice-borne diseases.[27] The writer, Professor Thomas, stressed that any method for the microbiological control of rabbits had to be sufficiently injurious to cause their death but not affect domestic animals. On balance, he viewed disease as a supplementary method of pest destruction. For their part, sheep owners favoured ferrets, stoats and weasels, and government arranged supplies of ferrets through district rabbit inspectors, who sold them to farmers for about 7s 6d a head. Their popularity is evident in the following sales figures for the period between 1 January 1887 and 8 June 1888: Invercargill (1712 ferrets from six sites), Lawrence (10,700 from four sites), Queenstown (270 from one site), Balclutha (4300 from two sites) and Kaikoura (4550 from one site).[28] In 1887

Men skinning rabbits, Tuapeka district, Otago, early twentieth century.
HOCKEN COLLECTIONS, UARE TAOKA O HĀKENA, UNIVERSITY OF OTAGO

an official report showed that 1.3 million acres of public land in rabbit-infested parts of southern New Zealand had been abandoned. Although twenty stations had been re-let at greatly reduced rentals, 0.4 million acres of former pastoral land remained unoccupied.[29]

Erected in the hope of halting the pest's northward spread, extensive rabbit-proof fencing proved little better than an expedient.[30] Once established in an area, rabbits quickly increased in number and competed with livestock for herbage, as the aptly named 'Leporicide' wrote in the 6 December 1879 issue of the *Otago Witness*:

Not only have the pastoral capabilities of a large portion of Southland and south-western Otago been almost ruined by the destruction of the herbage and the fouling of ground by the arid excrement of the rabbit, but the very scrub and sap-lings are everywhere barked and destroyed, nor has even that peculiarly tough morsel the spear-grass escaped. On the high country, the numerous valuable fodder plants and herbs, such as cotton plant, aniseed, blue tussock, &c. are being rapidly exterminated. Only the tutu is spared to slaughter the unfortunate stock who survive the universal starvation.

Rabbits continued to provide sport, food and pelts for rural settlers, but they spread and multiplied to become a plague of Old Testament proportions on pas-toral properties in the drier environments of inland South Island. There were insufficient native, and no specially introduced, birds of prey, so landowners had little choice but to exterminate rabbits if they wished to run sheep and cattle profitably on their properties. In the mid-1880s some rabbit inspectors were calling for total protection of weka, which by then had almost been eliminated from the lowlands of southern New Zealand, as they were showing promise as predators of young rabbits. Extermination could be prodigiously expensive, as the following outlays, compiled from James Preston's cash book for Black Forest Station for the ten months from 25 April 1916 to 28 February 1917, show:

Mustering	£68 15s
Shepherding	£48 15s
Shearing	£152 1s
Shearing & mustering	£51 5s
Poisoning rabbits	£22 10s
Rabbiting	£286 14s

Out of the total of £630 0s 6d, the listed expenditures on rabbit extermination accounted for 49 per cent.

Rabbit control was a perennial, time-consuming and expensive affair, and not something that could be achieved through two or three years of effort. It also called for commitment on the part of the men contracted to spread poi-soned bait over affected properties. On 20 June 1884 the manager of Aviemore Station wrote to the London backers saying he had been 'very busy laying poison for the rabbits . . . or rather supervising it myself because it is impossible to get hired hands to do it properly. If they are given, say, 20 pounds of poisoned oats to carry onto the hills, they get rid of it as quickly as they can.'[31] Over the twelve months to 1 May 1884, the wages of rabbiters employed on Preston's Aviemore Station in the Waitaki Valley came to £649, or 21 per cent of total annual expenditure, and when sold at auction the skins realised £425. The price received for rabbit skins depended upon time of year and condition of the pelt, and they were usually graded on the property before being sent to an agent in a nearby town or provincial centre for consignment to Great Britain. On 24 June 1916, J. Rattray & Son of Dunedin informed Preston that the two bales of rabbit skins from Haldon Station sold on his behalf had yielded £59 13s 9d, and the notice dissected the consignment of skins as follows:

Early autumn
Autumn
Early winter
Winter black
Spring
Milky doe
Summer
Fawn
Other black
Racks
Attacked by weevils
Hare skins

At Haldon Station, also operated by Preston and his family, the wage bill for rabbiters – shooters and trappers as well as poisoners – was £275 in 1904, £518 in 1905 and £339 in 1906, and these sums were by no means the full cost.

As tussock grass and low shrub ecosystems came under the plough for managed pastures of introduced grasses and broadleaf herbs, resident populations of rabbits declined to more readily manageable numbers. With the help of large outlays of money and labour on eradication, and the erection of rabbit-proof fences around fields, the occupier of a large block of land could bring rabbits under control, but this counted for little if his neighbours and the state were not equally assiduous. A five-wire fence along the property boundary might stop sheep from entering but was no barrier to rabbits and stray dogs, so settlers initially concentrated their rabbit eradication activities along vulnerable property boundaries. Some, like James Preston of Aviemore Station in the Mackenzie

Country, called on their neighbours to split the cost of erecting rabbit-proof fencing along the surveyed line between them. In a letter to Preston on 29 June 1903, William Grant of Grampian Station stated that he saw no need for a rabbit-proof fence between their properties, and asked Preston to set out his reasons before he would consent to its erection and commit to paying half the cost.[32] Eight years later the Timaru Branch of the Canterbury Farmers' Co-operative Association quoted Preston £30 for the rabbit-proof iron mesh and £10 for the metal strainers needed per mile of the planned ten miles of boundary fence.[33] When the unnamed occupant of an adjoining property refused to pay half the cost, Preston petitioned the Commissioner of Crown Lands in Christchurch and on 4 March 1912 was informed that the agency declined to interfere because this 'must be a matter for mutual consent'.[34]

Analysis of long runs of farm and station diaries allowed me to track the spread, establishment and rise of rabbit numbers on individual properties, and financial records showed landholders' investments in time and money to control if not exterminate the pest. Eradication was driven by the economic imperative to rid the property of a feral herbivore that competed with sheep and cattle for grass, damaged dry land turfs, and dug holes deep enough to trip horses and cattle. Government inspectors were appointed to visit properties, monitor the effectiveness of legislated rabbit control measures and impose penalties for non-compliance. As Archibald Morton wrote to Preston on 12 July 1898, 'What the Department requires is that the rabbits be destroyed, and effi-cient steps for that purpose must be taken. Under no circumstances must the rabbit be allowed to increase to anything like what they were last summer.'[35]

Methods of Eradication

Initially, shooting was the preferred method of rabbit control, and family mem-bers, shepherds and hired labourers carried firearms as they tended flocks of sheep and herds of cattle on higher ground. Shooting was effective for as long as rabbit densities remained low, but inadequate for complete control. Mass poi-soning was a more successful option even though landowners soon found that no one method was certain to work in successive years. During the first decade of the rabbit plague in inland South Island, cracked wheat dressed with white phosphorus was the usual bait but, according to 'Leporicide' in a letter published by the *Otago Witness* on 6 December 1877, it was not economically feasible to lay poisoned grain over a property larger than 50,000 acres. Furthermore, at

Rabbiters' huts, Stony Creek, Southland, late nineteenth or early twentieth century.
HOCKEN COLLECTIONS, UARE TAOKA O HĀKENA, UNIVERSITY OF OTAGO

least two bushels (approximately 70 litres) of phosphorised wheat were needed for every 200 acres (approximately 80 hectares) of heavily infested land. Even fumigation with carbon disulphide or carbon monoxide gas could impose a heavy fiscal burden on a landholder because all entrances had to be sealed with the aid of pick and shovel before the gas could be pumped into a warren.

Of the runs of farm and station diaries from the nineteenth century that I read, only two contained direct references to poisoning rabbits with carbon disulphide gas, and then for only one season, although several landholders used it to dissolve phosphorus when mixing rabbit poison. One was James Preston at Longlands Station who, on 20 September 1887, recorded obtaining 'Bi Sulphide of Carbon' and using it to poison rabbits in the vegetable garden. On 15 October he recorded 'shooting rabbits and bi-sulphiding the holes'. From that date onwards,

Preston and his employees laid poisoned bait and shot rabbits on the property.

Oats were cheaper than wheat and widely used for poisoned bait.[36] Irrespective of which grain was used, it had to be cracked rather than crushed to permit a greater coverage of phosphorus without compromising its attractiveness to rabbits. If the grain germinated, the bitter taste could deter rabbits from eating it. Addition of sugar – sometimes treacle, even fruit jam – increased the bait's palatability. On Aviemore Station, and presumably elsewhere in the semi-arid lands of the central South Island, the manager, William Shirres, often replaced sugar with salt in the recipe for phosphorised oats.[37] Oil of rhodium, a commercially available essential oil derived from a tropical American tree, was found to make the poisoned bait more attractive to rabbits and, like oil of aniseed, was occasionally added in small quantities to the cooled mixture. I found few references to rhodium in nineteenth-century farm and station diaries from southern New Zealand, but on 18 June 1881 the landholder of Orari Station, William Macdonald, placed the following order with a chemist in Christchurch: three pounds of phosphorus in bottles each containing a quarter pound of the substance, and three one-ounce bottles of rhodium.

White phosphorus is highly reactive and liable to ignite spontaneously when exposed to the air. It is also toxic to farm animals and people. For those reasons, it was normal practice for the property owner, a member of his family or a trusted employee to mix just enough poisoned bait to lay out over the following two days. White phosphorus was usually sold in sealed cans, brought on to the property in small amounts and used quickly. When not in use, the pots, ladles and barrels used to prepare and store poisoned bait were kept under damp sacks to guard against spontaneous combustion. It was the usual practice to visit bait stations over several days to check if the poison had been taken. Experience had shown rural people that winter was the best season for poisoning, and it had the further advantage of more valuable skins than those taken in summer and autumn. In addition, during the summer months, family members and employees were too busy mustering, shearing, culling, weaning and dipping, then bringing in the grain harvest, to attend to the rabbit problem. But at the height of the first rabbit plague in Central Otago phosphorised grain was laid down from February through November.

In the course of field trials, a correspondent of the *Otago Witness* on 6 December 1877 noted that the following mixture worked well. Dissolve four pounds of sugar in four gallons of water heated to 150° Fahrenheit, then stir in about one bushel of cracked wheat. At the same time bring to the boil one quart of water in a small cast-iron pot, add one pound of white phosphorus and continue boiling until the phosphorus has 'dissolved', then thoroughly stir it into the mixture of cracked wheat and sugar syrup. As it cooled this mixture gave off toxic fumes, and makers were advised to sprinkle on a layer of dried cracked wheat, let the mixture cool a little before resuming stirring, and repeat as many times as necessary until the fumes had abated and the poisoned grain was cool. The prepared bait was left to rest overnight under a cover of wet sacks before one fluid ounce of oil of rhodium was stirred into the mixture. Small amounts of unadulterated cracked grain were distributed in infested areas for several days to accustom rabbits to a novel food before the poison bait was laid on ploughed lines or well-used runways leading into warrens.

The correspondence between Archibald Morton, formerly the landholder then the manager of Haldon Station, and James Preston encapsulated many of the issues raised by rabbit control on pastoral properties in inland South Island during the late nineteenth century.[38] On 23 May 1894 he reported bringing in 70 bags of oats and one dozen ten-pound tins of phosphorus from a supplier in Timaru, and wrote to Preston that the annual rent for Haldon Station struck him as high in relation to the expenditure needed for rabbit control. Eighteen months later, on 24 September 1895, he mentioned that he had decided to experiment with phosphorised pollard because rabbits had begun to reject poisoned oats. A rabbit inspector who visited the station early the following year complimented Morton on his efforts, but the manager's report to Preston on 31 August 1896 contained mixed news. He was holding 13,000 good-quality rabbit skins ready for sale and expected that 'my poisoning will not have cost me much', but he had begun to find that the rabbits 'took the poison better towards spring than they did in May. It was the same with [poisoned] grain – they did not like it', and his neighbours were in the same boat. On 15 September 1896 he informed Preston that one of them 'had four men on [rabbiting] this last month but has given it up. I had Ted Cameron here week before last. He was saying most of the stations in [the] Waitaki [Valley] had given up on pollard for summer work.... [and] where people had been working with it last summer, the winter poisoning was a failure.'

Towards the end of the nineteenth century, phosphorised pollard was preferred over phosphorised oats or wheat because it was readily mixed, relatively stable, safer to store and had a longer field life. Rural people could buy 'poisoned sugar' in bulk,[39] but Preston's is the only reference I have found to this. Phosphorised pollard was also used to kill the flocks of small birds that fed on maturing crops of wheat and oats. In 1897 the New Zealand Department of Agriculture published two recipes for phosphorised pollard, and a well-used copy has been preserved in the Preston family papers. The first recipe called for the farmer to dissolve two sticks of white phosphorus (about four ounces) in a boiling hot solution of four-and-a-half pounds of brown sugar and three quarts of water, stir in sufficient pollard to make a stiff paste ready to knead, dust with dry pollard, then roll out and cut into small pieces ready to set out in an infested area. The second involved putting half a pint of cold water in a screw-top jar, adding one tablespoon of carbon disulphide and one-and-a-half sticks of white phosphorus, replacing the lid, and leaving the phosphorus to dissolve. This solution was added to three-and-a-half pints of hot water in which three pounds of sugar had been dissolved, then mixed with sufficient pollard to form a stiff paste. With both recipes, farmers were advised to use one-third less sugar and replace it weight-for-weight with treacle during very dry weather to ensure that the poisoned bait did not dry out. Settlers were advised to place small samples of the prepared bait on sods overturned with a spade or at intervals along a ploughed furrow, the amount depending on the density of the pest. They were also cautioned to remove livestock before laying out poisoned bait, to keep unused phosphorus covered by water and preferably well away from buildings, and to avoid build-up of toxic carbon disulphide fumes in confined spaces. This leaflet was published in a first edition of 10,600 copies in 1897 and a second printing of 25,000 copies in 1905. In its 8 May 1907 issue, the *Otago Witness* published the second of those recipes, virtually guaranteeing that it would be widely read throughout the southern South Island, and warned that livestock, including farm and station dogs, must be excluded from places where bait would be laid to ensure they were not poisoned.

The following recipe was handwritten at the front of the Cody family diary for 1915 and was clearly intended for a smaller property with few rabbits.[40] Farmers were advised to dissolve a half pannikin of sugar in a pannikin of boiling water, add one inch of white phosphorus cut from a two-ounce stick and stir until dissolved. They were then instructed to add a pannikin of pollard, stir until well mixed, and then add twenty drops of oil of aniseed to the cooled mixture. This amount was sufficient to fill a fourteen-inch golden syrup can, in which it could be stored until needed. Quantities could be doubled if required.

During the mid-twentieth century, phosphorised pollard was found suitable for use during summer, but carrots dressed with strychnine were better in winter.[41] (Efficient distribution of the latter required aeroplanes or suitable mechanised land transport.) For the remainder of the year, hunting with a dog, shooting or trapping were the preferred methods of eradication. McLean and his gang in the Wairarapa also tried calcium cyanide – which gave off a poisonous gas when wet – in late winter, as well as tear gas and carbon disulphide. After experimentation, they found that addition of quince jam, which was expensive, or the cheaper apple and raspberry jam, even raspberry essence, made the bait more attractive and less likely than oil of aniseed to attract sheep and cattle.

McLean noted that cats, weasels, stoats and ferrets were widely liberated across New Zealand during the 1880s 'in the hope that they would help to control the hordes of rabbits that had begun to ravage the land'.[42] Wild cats tended to wait for rabbits to emerge from a burrow and, like mustelids, enjoyed protection under the Rabbit Nuisance Act 1886 outside built-up areas. In the Wairarapa, McLean found that 'Ferrets, stoats, and weasels require a high population of rabbits if they are to survive in any numbers. With low rabbit densities they quickly die out . . . . as soon as the rabbit population fell below a certain level the cats and ferrets began to die out, and from then on we seem to have had a plague of rabbits every ten years or so.'[43]

The Rabbit Plague at Sunnyside

The nineteenth-century farm diaries and letter books from southern New Zealand that I read usually mentioned rabbits, but John Wither's diaries proved invaluable for their documentation of a single landholder's experience over several decades on the one property – in this case, Sunnyside on the southwestern shore of Lake Wakatipu – starting in August 1871. Areas of moist friable soil on valley floors or on the shores of the lake were cleared of their native plant cover, cultivated, then laid down in wheat, oats, turnips and potatoes. Introduced animal and economic or decorative plant species were generally ascribed high value, although rabbits caused considerable problems within a year or two of reaching the property and were soon to tie up considerable resources of time and energy, materials and money.

There can be no claim of completeness, and Wither was not consistent in reporting how many rabbits he and his employees killed, but his diary entries tell us much about a critically important period in the spread, establishment and growth of rabbit populations in the dry country of Central Otago. His first references to rabbits were in July 1877 when on three days he reported that some had been shot on the property (Figure 6.1, opposite). He next mentioned rabbits in September, followed by a gap until March 1878. From then until the end of November there was at least one reference per month to a family member or a hired hand shooting rabbits on the property. Much the same was true in 1879, although there were more frequent references to shooting and there was the first mention of 'poisoning' rabbits. What had been a slow increase in the pace of eradication over the three years from 1877 accelerated in 1880, when poison was laid each month from March to December, and there was some hunting with guns, usually with the aid of dogs, for six of the ten months. In 1881, Wither first reported buying traps with which to catch rabbits.

The usual bait was phosphorised wheat, which Wither and his men laid in winter when rabbits had little to eat because the herbage had been browsed to the ground in heavily infested areas or palatable shoots were covered by ice and snow. Like all farmers who spread poisoned grain, Wither found that rabbits did not take the same bait from one year to the next, so he tried several different baits. On one occasion (10 April 1880) he reported mixing 'rodium' into the slurry of phosphorised grain, but did not comment on its success and made no further reference to it in his diary. On 19 October 1881 he tried 'bisulphate of carbon' (*sic*) that he and his men could have pumped into rabbit burrows or added to the phosphorised grain. Several later entries referred to 'putting poison in holes' (for example, 11, 18 and 19 November 1881; 20, 21 and 24 April; and 7 November 1882), which may signify the use of carbon disulphide because Wither's usual term for setting out poisoned grain was 'laying poison', but he did not record the success of this alternative method of pest destruction. He also tried trapping and catching rabbits with ferrets, and this sequence of quotations from his diary – the first to mention ferrets – for late 1882 is chilling to twenty-first century eyes:

Poisoned some rabbit holes, and ferreted some rabbits. Got a ferret from
Mr Hassel. (17 November)
Killed some rabbits with ferreting. (18 November)
Repairing fences and ferreting. (22 November)
Two ferrets lost . . . . weeding in garden and ferreting. (28 November)
Looking after [for?] ferrets. One ferret found. (30 November)

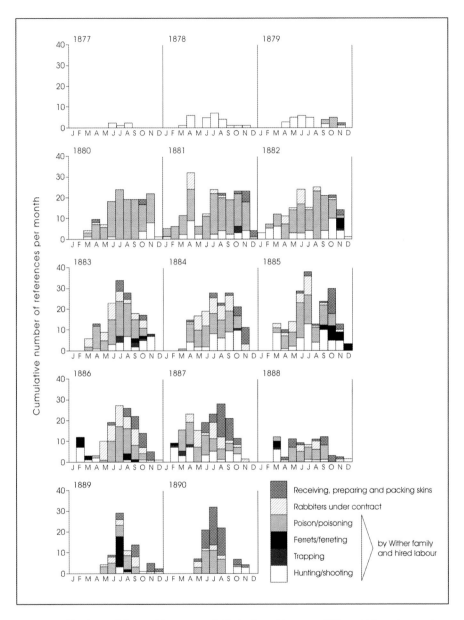

FIGURE 6.1 Numbers of diary entries per month that referred to ways of killing rabbits and preparing their skins for sale at John Wither's property, Sunnyside, in the Lake Wakatipu basin from 1877 to 1890.

SOURCE: INFORMATION COLLATED FROM JOHN WITHER'S DIARIES; SEE APPENDIX NOTE 8

The evidently small involvement of Wither and his hired hands in rabbit eradication during summer and early autumn reflected the failure of poisoning programmes during seasons when more attractive supplies of food were available to rabbits, but it was also because Wither, his family members and employees were fully engaged with mustering, shearing and culling sheep, followed by wool scouring, then the grain and potato harvests. When inclement weather halted shearing, the men were sometimes deployed shooting rabbits on the property, but rabbit eradication tended to be a task for the cool and cold seasons. Figure 6.1 shows that between 1879 and 1890, Wither had progressively to employ more diversified methods for rabbit eradication, including a lesser reliance on poisoning and an increase in trapping and shooting.

Wither's diaries do not tell us with uniform precision how many men were involved each day with rabbit extermination, but his diary entries implied that there was some such activity most days during the winter months of 1881 and that it involved as many as five people – normally two or three of them employed under contract – at its seasonal peak. I was unable to calculate the actual cost to Wither of rabbit eradication, or even the total value of clean dry rabbit skins sent off the property for sale, but the following numbers are indicative. In 1880, the first year of intensive rabbit poisoning at Sunnyside, Wither distributed 28 pounds of poisoned grain. The following year, the total was in excess of 1900 pounds. In 1881 at least 3400 pounds were laid, and in 1882 the total was about 3600 pounds. In his diary he recorded sending 575 rabbit skins to Invercargill for sale in November 1879, 1920 skins in October 1880 at a reserve price of 1s 7d each, and three bales of cleaned and dried skins in November 1881. In winter and spring, Wither hired professional rabbit shooters for between 10s and 12s 6d per week plus a payment of 1d or 2d per skin. His income from the sale of rabbit skins appears modest in relation to the recorded costs of labour and materials, let alone the loss of palatable herbage and diversion of labour from fencing, pasture improvement, drainage and other important development tasks.

Rabbit Extermination on James Preston's Properties

The Preston family leased and owned properties in the southeastern South Island, and James Preston and his employees moved frequently between them. His financial backers, agents and suppliers operated separate accounts for the properties, and many of those records have been preserved. Wherever he was living at the time, Preston kept detailed diaries and maintained a record of his financial transactions, including payments for specified materials and tasks performed by contract and other labour between 1890 and 1924. This information allowed me to estimate annual expenditures on rabbit control – notably for wages, ammunition, phosphorus, pollard, traps and ferrets (Figure 6.2) – and clear patterns are evident in the graphed numbers. On Ben Ohau and Haldon stations in the Mackenzie Country, annual expenditures on rabbit

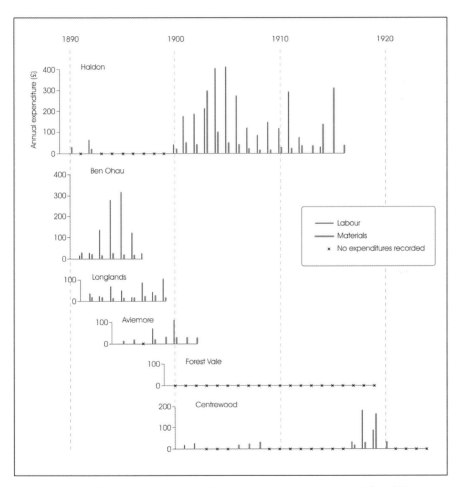

FIGURE 6.2 Reported expenditures in £ per annum on labour and materials for rabbit extermination on six rural properties in southern New Zealand operated by members of the Preston family for various periods between 1890 and 1924. SOURCE: INFORMATION COLLATED FROM JAMES PRESTON'S DIARIES AND PERSONAL PAPERS; SEE APPENDIX NOTE 8

extermination followed an almost decadal cycle, rising from less than £100 to more than £500 then falling away. On Longlands Station in the Maniototo, where Preston had been exterminating rabbits since the early 1880s, annual expenditures ranged between £50 and £125, and the outlays at Aviemore Station in the upper Waitaki Valley were similar. Forest Vale and Centrewood stations, in cooler and moister parts of the southeastern South Island, required only small annual expenditures on rabbit control. Most years, the cost of labour exceeded that of materials, although during the Great War, when agricultural labour was difficult to find and costly, poisoning was the preferred method of rabbit control at Haldon and Centrewood stations.

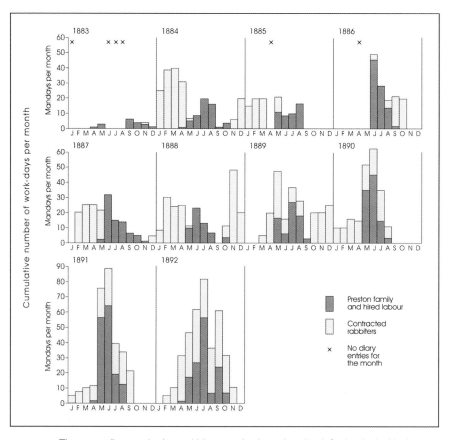

FIGURE 6.3 Time spent (in man-days) on rabbit extermination at Longlands Station in the Maniototo by (1) family members and hired labour, and (2) contract rabbiters between 1883 and 1892.
SOURCE: INFORMATION COLLATED FROM JAMES PRESTON'S DIARIES, SEE APPENDIX NOTE 8

Of the six sets of records, that for Longlands Station was the longest. Rabbit extermination began in earnest during 1883 and picked up during the succeeding years. Preston was in residence for the next eight years, and his diary entries were sufficiently detailed to permit fair estimates of numbers of man-days per month on rabbit extermination by family members, hired labourers and rabbiters working under contract (Figure 6.3).

Figure 6.4 collates references to the different extermination methods employed by Preston and his staff during each of the eight years from 1883, and suggests an initial period of experimentation followed by a steady shift to reliance on contract rabbiters, with family members and workers providing support by laying poisoned bait in infested areas.

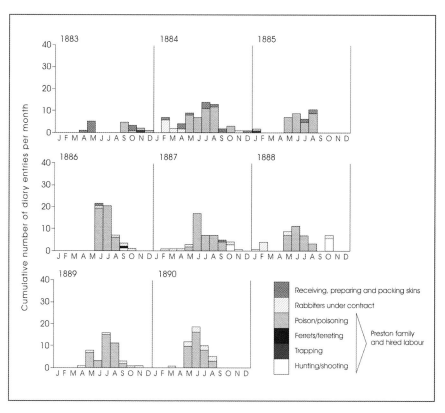

FIGURE 6.4 Numbers of entries per month that referred to the different modes of killing rabbits employed at Longlands Station between 1883 and 1890.

SOURCE: INFORMATION COLLATED FROM JAMES PRESTON'S DIARIES; SEE APPENDIX NOTE 8

Wither and Preston were flexible in their selection of methods for rabbit extermination on their respective properties. It was a costly business and there were no short-cuts. Like other landholders at the time, if they wished to control the pest they had to throw resources at the task and opt for the method best suited to rabbit densities, the annual schedule of work on the land and environmental conditions at the time. Flexibility also extended to the contractual arrangements that landholders such as Preston negotiated with professional rabbiters (Table 6.1).

TABLE 6.1 Terms of employment for rabbiters working under contract on Longlands Station at various times between 1884 and 1899.

Abraham Dilworth commences work this morning [at] 20/- a week. He finds himself and gets all the skins. (24 November 1884)
C. McAra began work at 30/- a week and finds himself. (8 February 1887)
Jones, rabbiter, keeps [i.e. patrols] the Kyeburn boundary from today at 15/- a week. (15 February 1888)
James Mann began work this morning. Rabbiting at 20/- a week and finds himself in everything and keeps his own skins. (24 October 1888)
E. Coglin paid me ten pounds ([£]6 cash, [£]4 cheque) for sole right [to] collect poisoned rabbit skins this season on almost 18,000 acres south and west of Swinburn [Peak]. (20 May 1889)
Ford the rabbiter began work this morning at 20/- a week. Finds himself and keeps his skins. (7 October 1889)
W. Creighton started rabbiting this morning at 20/- per week and finds himself. (25 November 1889)
Sold Ford and Crutchly the right to collect all poisoned rabbit skins this winter for the sum of twelve pounds, and they to poison a week for nothing. (13 May 1890)
McClusky and his mate started trapping in the Houndburn at 15/- each, find themselves and keep their skins. (19 September 1892)
Baldwin started to trap this morning at 20/- a week, find himself and keep his skins. He got 110 traps [from me] for which he pays eleven pence for all he loses. (5 October 1892)
Gave Jimmy Fletcher poison for Houndburn paddocks, which he poisons for the skins. (12 June 1899)
Two trappers out to Shag [River]: 35/- a week for the two and find; [received] 12/- for big and 6/- for small skins per 100. (29 October 1899)

SOURCE: INFORMATION COLLATED FROM JAMES PRESTON'S DIARIES, SEE APPENDIX NOTE 8

Rabbit Extermination Elsewhere in Southern New Zealand

In his monthly letters to the absentee landholder, the manager of Ida Valley Station in the dry country of Central Otago showed an observant pragmatist's understanding of the causes and consequences of fluctuating numbers of rabbits. On 24 April 1881 he wrote that the combination of a hard winter, a major shortfall in the water supply and an abundance of rabbits meant that the sheep would 'have a hard struggle' on the property. In December he reported that 'our rabbits are returning and breeding at a fearful rate' and was clearly desirous of finding an effective regime to exterminate them. Like Wither in the Wakatipu basin, he laid poison during winter even though a heavy snowfall could 'tell seriously on the expenses of the [rabbiting] gangs', and on 30 April of the following year he justified his regime of rabbit poisoning to the absentee landholder:

> Your letter implies, first, that summer killing of rabbits is undesirable. Second, that the work has been done more expensively than it might have been . . . . Unfortunately, the question of whether or not this cost is more than returned by the extra stock one can keep, and the amount of wool one can get from them, is one that cannot be answered except by experiment, and I regret having to try it. My own experience is very decided, and it is confirmed by the experience of anyone I know who has tried it. There may be a middle course between not killing in summer and killing as many as possible, and we should have to try to find it.

The task of rabbit eradication was a significant challenge to the manager, and on 2 June 1892 he wrote again to the absentee landholder that 'the weather so far has been bad for poisoning, an odd frosty night and then rain threatening snow'. One month later he reported 'the weather has been very much against successful work so far, and the rabbits have not been taking the bait well. If frost should set in after this last burst [of very cold weather] I think we shall do better.' One month after that he believed 'if there should be an early spring (as seems likely) we may be in difficulties before we get over all the ground [laying poison bait]. But it's a striking commentary on the strife this cursed [rabbit] plague has got on us.' His skill as an observer is evident in the following three statements from his reports to the absentee landholder:

> There has been no snow, sometimes the frost has been so severe and my experience has been that during the hard frosts the rabbits will not take the poison well. (12 July 1894)

The weather since I last wrote has been 'good', except for a slight fall of snow yesterday and Saturday, which will do no harm and may kill some of the young rabbits. (3 September 1894)

The snow killed many rabbits on the higher ground, and except in patches they are not very numerous now. (1 October 1895)

His optimism proved short-lived, and on 5 November 1895 he informed the absentee landholder that 'the rabbits are swarming in places, making me very nervous and anxious'. His efforts to control the plague were redoubled during the remaining years of the nineteenth century, but the setbacks of earlier times recurred. On 24 June 1896 he wrote in the station diary that he had seen a 'good many rabbits, but [there] have been quite as many in other years after poisoning'. A month later he reported being 'not at all satisfied with results of poisoning – plenty of poison layed but [there are] great number[s] of rabbits on Poolburn face and elsewhere [on the property]'. Three days after that, on 11 August 1895, he borrowed two ten-pound tins of phosphorus from nearby Matakanui Station to mix with grain and lay in the worst affected areas of the property. In the last extant diary, he recorded on 3 March 1899 that he 'did not see many rabbits [on the property] . . . the poisons (*sic*) seemed to be clearing them well'.

David Bryce farmed between Milton and Balclutha in south Otago, and the first reference to rabbits in his diaries – the extant copies of which run from September 1873 to the early years of the twentieth century – was dated 3 June 1886, when 'the rabbiters got 48 rabbits'. The following year (26 May 1887), he reported shooting a rabbit. Four days later he killed eleven rabbits and found two poisoned animals. A rabbit inspector visited the farm on 12 December 1888, but Bryce did not record the outcome. The pace of rabbit extermination apparently accelerated during the last five years of the nineteenth century, involving most family members and at least one rabbiter hired under contract as well as distribution of poisoned grain. On 2 February 1895 he recorded that 'McLeod is getting a good few rabbits', and on 13 April 'the boys are rabbiting [and] got 95 with the dogs'. The following month 'Tommy got 30 rabbits'. The implication of these few references is that during the late nineteenth century rabbits might have been an annoyance for Bryce but they were not the severe threat to economic viability that they were on Wither's property in the Lake Wakatipu basin.

Although the documentary record is incomplete, much the same seems to have been true of the McMaster family property at Tokarahi in the lower Waitaki Valley, and there were no references to rabbits in the four extant diaries between 1865 and 1893. In northern Southland, Joseph Davidson recorded on 12 November 1885 the adverse effects of caterpillars, grass grub and small birds attacking grain and pasture plants: 'The birds is (*sic*) playing havoc with my oats – pulling the blades up and eating the seed. I never saw them do the same before.' But the only references to rabbits in his diaries were on 16 February 1885 when he described skewering dead rabbits then placing the stakes around his crops to attract hawks to 'frighten the birds from eating grain', and on 12 November 1891 when flooding caused 'a great slaughter among the rabbits' along the banks of the Waikaia River. Other landholders also found that a wet winter with widespread flooding led to many rabbits drowning in burrows dug into riverbanks, whereas a dry winter usually resulted in a major increase in numbers the following spring.

Across Southland, farmers set rabbit traps in the afternoon and collected the carcasses early the following morning. The animals were skinned, gutted and cleaned, then placed in roadside safes ready to be collected by one of the many freezing-works trucks that travelled every day of the week throughout Southland. McLean wrote, 'We've paid our way during the hard times over the past fifteen years with money from the sale of rabbit-skins and carcasses, and it is the rabbit that has kept us on our farms.'[44]

In its edition dated 27 May 1887, the *Weekly Times* published a brief report from its correspondent in Woodlands, a small community west of Invercargill, which included these words: 'The Meat Processing Works are still going on preserving mutton and rabbits, this gives employment to a goodly number of men, and long may it continue.' Much the same was true of the Cody family, Irish immigrants who farmed at Lime Hills on the floodplain of the Oreti River in central Southland. Between mid-1901 and the end of 1916, albeit with gaps in the extant record, they described several episodes of rabbiting. The first, on 17 June 1901, recorded them receiving £1 17s 9d for rabbit skins and a [cattle?] hide. Several references to poisoning rabbits and collecting their skins followed. On 19 October 1901 the family received £2 for rabbit skins, and a further £2 5s three months later. Between mid-February and the end of April 1902, they sent away 650 rabbit skins in ten bags to one merchant and 275 skins in six bags to another. The animals had been trapped or poisoned on the Cody family farm as well as on neighbouring properties. On 15 March 1908, Cody ordered 'four coils of the strongest rabbit netting', and on 27 January 1910 he received a visit from a rabbit inspector – he did not record the reason or note the official findings – and a follow-up visit seven days later. Family members continued to trap and shoot rabbits, and Cody senior occasionally employed a rabbiter under contract for a few weeks, but the regime had changed from one of hunting rabbits for income to one of controlling their numbers by installing rabbit-proof fencing around areas of persistent infestation, and once or twice during spring or early summer laying down poisoned bait in any remaining areas of rough ground on the low hill country where rabbits could still find a congenial habitat.

The last of these accounts of rabbit extermination is based on the Ramsay family diaries for their small property near Hyde in the Maniototo, and covers the period from January 1911 until mid-1914 (Figure 6.5).[45] After a slow start in 1911, their efforts redoubled in the following two years. Young members of the

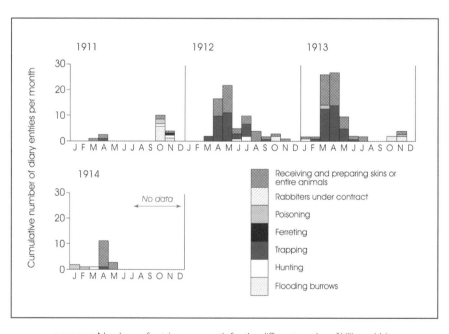

FIGURE 6.5 Numbers of entries per month for the different modes of killing rabbits to control numbers and for income on the Ramsay family farm near Hyde, on the fringes of Central Otago, between January 1911 and May 1914. SOURCE: INFORMATION COLLATED FROM THE RAMSAY FAMILY DIARIES; SEE NOTE 57

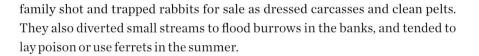

family shot and trapped rabbits for sale as dressed carcasses and clean pelts. They also diverted small streams to flood burrows in the banks, and tended to lay poison or use ferrets in the summer.

HERBACEOUS WEEDS

A useful definition of a weed is a plant growing in the wrong place, and we tend to think of weeds as introduced species like dandelion, nasella tussock and wandering Jew. Alfred Cockayne distinguished between migratory and long-duration weeds, the former persisting until the effects of disturbance cease, then dying out through competition with other plants, whereas the latter normally require human intervention to eradicate them.[46] Introduced weedy plant species were seldom noted in early farm diaries, but references to them were more frequent as land was brought under the plough for fields of potatoes, oats to feed the horses and wheat for sale to millers nearby. Given the sparse information available, one cannot be certain about this, but in the first two or three decades of organised settlement there seem to have been relatively few introduced weed species on rural properties, with only one or two of them dominating. Residents of properties with gorse hedges experienced problems with weeds when trying to get gorse seedlings established, as the manager of a property in north Otago, Nugent Wade, informed W. H. Teschemaker on 15 November 1875: 'The [gorse] hedges which have been planted this year are all doing well so far [but] they take a lot of clearing. They are sometimes smothered with weeds in some parts and we dare not touch them for fear of pulling up the [gorse] plants. We must let them grow before we can weed them.'[47] In the same letter, however, he also wrote, 'We have not yet finished trimming the gorse hedges here. It is quite a Herculean labour. It is a great pity that they should have been left untrimmed all these years as it has made them very stalky and weak.' The diaries for Kauru Hill and Taipo ran from 1864 to 1895, with breaks in 1885–86 and 1891–94, and showed annual references to thistles peaking in 1864, 1867–68 and 1878–79 when gorse hedges were being established on ploughed strips cut through tussock grassland, and native grassy turfs were being broken up for improved pasture, potatoes and winter root crops.[48] As on most lowland properties, thistles followed the plough but became less of a problem as soil fertility was depleted and farming operations matured.

Weeds usually entered a rural property as seeds caught up in the fleeces and pelts of livestock from other farms and stations, but the prime source was pollutants in sacks of seed purchased from seed merchants for establishing new pastures or sowing between burned tussocks. The seriousness of this became better known when the Department of Agriculture released the findings of its regular analyses of the purity of commercially available seed. Alfred Cockayne drew on this work for a widely distributed publication in which he sought to convince farmers that cheap impure seed was more costly in the long run than expensive clean seed.[49] He noted that farmers were required by government inspectors to control noxious weeds on their properties, and many did so, but argued for a wider definition of a noxious weed as one 'whose growth interferes with the yield of the crop or pasture'. Several of his conclusions must have worried pastoral farmers, notably that 'the vast majority of our weeds owe their origins to having been actually sown, in most cases quite unintentionally, along with ordinary agricultural seeds', and that more than 50 species of weedy plant were represented in the analysed seed samples. He referred to samples of commercial white clover seeds containing 36 different weedy plants, of red clover with 23 and of timothy with seven. Four years later, Cockayne published a summary of the findings of thousands of analyses conducted by the Department of Agriculture into impurities in commercially available seed for establishing pastures.[50] This must have greatly concerned landholders even though a warning had been published twelve years earlier in the widely read *New Zealand Farmer*.[51]

During the second half of the nineteenth century, thistles became a major nuisance throughout New Zealand. S. Grant and J. S. Foster reported thistledown being blown two miles out to sea in Hawke's Bay and landing on the deck of their steamer. They also described thistledown whitening gorse hedges at Waitotara in Taranaki and blowing 'across the streets of Christchurch like a slight snow storm'.[52] A petition was presented to government by residents of Canterbury relating to the Thistle Ordinance that had been introduced by the provincial council.[53] The nub of their argument was that while individuals were expected to keep control over thistles on their land, little was being done to control them on adjoining Crown land, Māori land or roadsides, which the petitioners described as 'thistle nurseries'. On 2 August 1876, Nugent Wade wrote to the absentee landholder, 'You can imagine what the thistles were like this year. We have some splendid hedges of them',[54] a theme he was to return to many times in his correspondence. On 18 October 1876 he wrote this prescient account: 'I foresee the same difficulties this year in connection with gorse fences. The gorse seed germinates so very slowly and the millions of thistles and other weeds come up so rapidly that the ground where the gorse seed has been sown is a dense carpet [of thistles] before the gorse appears, and one does not weed it or one pulls up all the gorse.'[55] He suspected pollutants in the grass seed then being brought onto the property as the prime source of his difficulties.

John Wither's daily records show that the perception, role and significance of native and introduced weedy plants changed over time. The references to 'thistles' – *Cirsium arvense*, *Cirsium vulgare* and *Onopordium acanthium*, it is impossible to state which – in his diaries steadily increased in number. The first entry to mention thistles was in 1876, and there were 21 entries eleven years later. During the summer, when employees could be released from mustering sheep, supporting the shearers, helping to scour the wool clip or working on the harvest, they chipped thistles in cultivated ground. There were other weedy plants, one of which, sheep sorrel, grew abundantly in Wither's fields of potatoes and turnips, but thistles received the greatest attention.

At Haldon Station, James Preston was also bothered by thistles and dock, which his men chipped with a sharp spade in midsummer, and a weedy 'geranium' that cannot be identified. Fat hen in cultivated land was a persistent source of difficulty for the manager of Ida Valley Station in Central Otago, and on 5 November 1891 he wrote to the absentee landholder, 'The fat hen has got such a hold over all ground here that it would be a folly to risk sowing turnips, even if a wet summer ensued.' This was the first reference to a weedy plant on that property in a bound collection of letters, the first of which is dated 24 December 1870. Throughout southern New Zealand, sheep sorrel was common on cultivated land, and William Macdonald of Orari Station mentioned it in his diary in October 1863.[56] Three decades later, Davidson was also plagued by sheep sorrel in his potato and turnip fields, and on 11 November 1889 wrote 'day very close and hot, sorrel growing quickly'. At that time, sorrel and other herbaceous weeds were usually cleared by grubbing, a task that required a considerable amount of manual labour. On 12 January 1891, Davidson recorded: 'Alek scuffling among the turnips – they are getting very bad with sorrel.' Three years later, on 4 January 1894, he wrote, 'day fine, weather looks steadier. I am afraid [the potato crop] is going to be a dead failure here this year. The frost seems to have gone into the roots of some, and the sorrel is growing very rank among them that it makes one continually hoeing to give the potatoes a start again after the frost [on 16 December].' Four days later he recorded that 'the sorrel is growing very fast – I don't know that I ever saw the potatoes as backward as they are this year. The [recent] frost seems to have gone to the roots of them.'

For its part, the Ramsay family at Hyde in the Maniototo was bothered by California thistle and 'yellow weed' (possibly ragwort).[57] After an absence of references to weedy plants between 1886 and 1895, on 16 December 1895 David Bryce mentioned that 'Tommy [was] up on the rough ground cutting some [unnamed] weeds.' From then until the middle of 1900, thistles were the only weeds identified by name in his diaries. The Cody family of Lime Hills in northern Southland did not refer to native or introduced weedy plants in their diaries between 1901 and 1906, but each year thereafter California and possibly Scotch thistles required between five and twenty man-days during summer and early autumn to grub them out by hand or mow.[58] James Miller's references to weeds on his small farm in the low hill country of eastern Otago over the eight years to the end of 1910 were primarily to California thistle, with occasional reports of dock and an unidentifiable 'weed'.[59] The letter books of the New Zealand Australian Land Company contained a copy of a letter dated 4 February 1870 reporting a manager's wish to conduct trials on a south Canterbury property with sow thistle, a potential substitute for rape as a supplementary feed source for livestock. He requested permission to hire someone to help him collect sufficient seed for a small-scale trial.[60]

NOVEL ECOSYSTEMS, NEW CHALLENGES

Vegetation clearance, soil erosion, declining soil fertility, rabbits and weeds were closely entwined during the second half of the nineteenth century. On many lowland properties, thistles and other weeds followed the plough. Their densities peaked while the rural landscape was undergoing transformation and began to come under control as farming operations matured. Many of the problems that pioneer families experienced with weedy plants stemmed from the short-lived fertility of the surface soil. As in the grasslands of the United States and Canada, the plough that broke the plain in southern New Zealand exposed a topsoil with plentiful organic matter that was able to retain rain water and snowmelt for normal plant growth, and an abundance of plant nutrients that had accumulated over several centuries. For a decade or longer, the soil supported successive crops of grain and pasture plants. But the bounty was finite, and by the 1880s landholders were seeing the economic consequences of a depleted soil nutrient pool: reduced carrying capacity, smaller wool clips, lighter grain harvests and increasing difficulty in managing sown pastures. Clearance of the area's tussock and low shrublands opened the way for pest plants and animals to become established, and absence of specialised animal browsers and predators in the New Zealand biota gave many of them free rein. At that time, biological controls and targeted herbicides were unavailable, so landholders had to apply costly, labour-intensive methods if they were to restrain the density and distribution of plant and animal pests, and they learned that modified ecosystems raise novel challenges that require painstaking management. Late in the second half of the nineteenth century, settlers were introduced to the importance of living ecologically, and many realised the economic costs and environmental consequences of not doing so.

CHAPTER SEVEN

Opportunities to See, Hear and Compare Meetings, Sales, Competitions and Exhibitions

In southern New Zealand, their farm or station was a settler family's anchorage. Each able-bodied resident had a roster of tasks on the property and made occasional visits to regional centres or nearby towns. Trips off the property enabled members of landholding families to meet their neighbours, compare notes about the effects of recent spells of unusual weather, hear about wool clips and animal growth and, more importantly, assist in the development of a community spirit in the district. The mixture of social and economic activities involved participation in or attendance at such competitive activities as wintertime ploughing matches, agricultural and pastoral association shows, and seasonal exhibitions mounted by horticultural societies. As their operations developed, settlers attended sales where livestock and sometimes neighbouring properties were put up for auction. These events were especially important when a landholder wanted to sell surplus livestock, replace animals that had died during a spell of unusual weather, seek replacements for animals sold to stock buyers, or buy a piece of farm machinery, but they also kept a landholder abreast of what his neighbours were achieving and the market was interested in buying. Other activities off the property were more social than economic. They helped a family set down roots in the district, ensured development of the local community, and included membership of road and drainage boards, raising money to support the poor and elderly, serving on church, school and hospital committees, and helping their neighbours organise sports events, dances and community picnics. These diverse economic and social opportunities are discussed in this chapter.

Attendance at church services provided invaluable weekly opportunities for rural people to meet their neighbours, inquire about their well-being, and exchange news of family and the district. Even so, it was difficult to quantify church attendance from entries in settlers' diaries. Sundays were normally rest days: times for reading, writing letters, cleaning one's living quarters, washing and mending work clothes, and entertaining visitors. A few *pro forma* farm and station diaries left sufficient room for a resident to enter brief notes about visitors welcomed and visits made, but many did not. Even where there was space on the page it was often left blank. As a consequence, one cannot be certain about Sunday activities, although entries for the other six days of a week frequently implied that church attendance was normal for families living close to settlements but not necessarily for their hired labour. Each year, isolated farms and stations were visited by ordained ministers who led services of worship, conducted baptisms and comforted seriously ill family members, employees and neighbours. Observance of Easter and Christmas was variable, and diary entries often suggested that for most farm and station families, the two Christian festivals were simply noted. In contrast, New Year's Eve and the following day were occasions for relaxation, attending sports meetings and horse races, going on community and family picnics, and carousing.

Of those individuals in Otago and Southland whose diaries and letter books I read, most were Scots Presbyterian who relaxed their observance of Sunday as a day of rest only when harvesting or shearing had not progressed as well as expected, usually because of strong wind, rain or persistently damp weather. The men might then spend the day mustering flocks off the slopes and into the holding yards, or putting as many animals as possible under cover in readiness for shearing the following day. Similarly, when a severe winter storm with heavy snow threatened the survival of sheep on higher ground, male family members and employees would spend the day raking them out of the snow, often with assistance from their neighbours, and guiding the sheep to bales of hay or fields of turnips on the flats close to the homestead. Before mechanised transport became widely available, it was easier to take animals to supplementary feed than *vice versa.*

Rural people were good observers of environmental conditions and were seldom isolated for long from information about the outside world. Most of the individuals whose diaries I read were also active in the economic and social development of their respective districts and welcomed opportunities to compare their livestock and crop yields with those of other landholders, then to use that information to improve the operation and profitability of their own properties.

GARDENING AND HORTICULTURE SHOWS

A strong competitive spirit was evident in many aspects of public life during the second half of the nineteenth century, including exhibitions organised by the horticultural societies that sprang up across the country.[1] A well-stocked and attractively laid out garden with high yielding fruits and vegetables showed the owner's standing as a family provider, and newspapers regularly published columns advising settlers when to divide, prune, graft and sow. The 14 June 1862 issue of the *Otago Witness*, for example, informed its readers how to establish tomato plants for early planting: fill a hollowed-out turnip with friable soil, sow several tomato seeds, keep the turnip and its contents warm and moist, plant out when the late spring weather is reliable, and train the growing shoots up a trellis. Six years later, in its 'Otago Farmers' Calendar' for the month of June, the same newspaper offered advice about when to till arable soil, sow wheat, plant rooted hawthorn cuttings and drain wet areas, as well as when seasonal work had to be done in orchards and vegetable gardens. The garden column in the 1 August 1868 issue of the *Otago Witness* advised readers that it was time to sow hardy flowering plants like lupin and sweetpea, plant out rooted cuttings of fuchsia and carnation, and prune or graft roses.[2]

As early as 1869, *The Press* in its 24 November issue was lauding gardening as an activity that would keep men away from public houses, feed their families, and ensure a congenial home environment for their wives and children. A private garden was a measure of the occupant's social standing, and owners of large properties in towns and rural areas hired professional gardeners to develop and tend orchards, vegetable and flower gardens. Such men could be trusted to make the best of a windy situation and exploit the benefits of different microclimates around the homestead. Landholders placed advertisements for individuals interested in securing employment as gardeners, and newspapers published lists of names and occupations of the passengers on recently arrived migrant ships, which helped the recruitment process.

Individual settlers corresponded with family members about garden, crop and orchard plants, and on 4 June 1864 *The Press* carried an advertisement placed by W. Hislop for his current catalogue of plants and general nursery stock. Six weeks later, H. E. Alport advertised an auction on 19 July at which 'new trees and shrubs & plants; fruit trees and bushes; variegated hollies, laurels and other rare shrubs; new roses, fuchsias, geraniums &c' would be sold. He advertised the availability of a catalogue in the 7 July edition of *The Press*, and the following July advertised a forthcoming auction of plants, including 'rare and valuable pot plants, in a consignment from Melbourne'. A month later, on 11 August 1865, Hislop was promoting a long list of plants for sale at his warehouse, including 120 named species and varieties of decorative plants.

My colleague Vaughan Wood and I compiled one list of decorative plants cited in published reports of the Christchurch Horticultural Society exhibitions and a second from those named in advertisements placed in *The Press* by local seed and plant merchants. We then compared the two lists with entries in a book by L. de Bray about decorative plants grown in Britain during the nineteenth century.[3] The latter included information about commonly planted species and varieties in 263 genera, of which 20 per cent were also listed in published reports of the Horticultural Society's exhibitions. In total, at least 30 per cent of the genera noted by de Bray were available commercially in Christchurch during the 1860s and listed in the society's published reports of its exhibitions during that decade. Of the 160 plant genera noted in those reports, 61 were described by de Bray. Twenty of the remaining 99 genera were native to New Zealand and 79 had been introduced. The native plants were predominately ferns (twelve genera), with five genera of flowering vines and small shrubs in second place. The taxonomical diversity of introduced plants is remarkable: 41 genera of flowering vines and shrubs; seven of tall shrubs, low trees and palms; seven of succulent plants and epiphytes; six of herbaceous plants; five of ferns and fern allies; and three of grasses and grass-like plants. Native plants appear to have enjoyed considerable curiosity value, and in their diaries early settlers occasionally referred to massed displays of ferns, leaves and flowering shoots of flax and toetoe to decorate public halls for dances, banquets and concerts. The remarkable diversity of native and introduced plants involved with gardens, exhibitions and public events is in marked contrast to the simplified species composition of land then being cleared of its native vegetation and converted to pastures or cultivated fields.

Large and small towns had horticultural societies. As well as shows and competitions, they mounted special exhibitions throughout the year to mark official visits to the town, even to demonstrate residents' allegiance to the monarchy,[4] thereby providing opportunities for town and country people to exhibit the best of their home-grown produce. One such display was described by *The Press* on 10 October 1864 as evidence of 'the rapid strides that the province [Canterbury] has lately made in the cultivation of European plants and flowers, and the interest that so many [residents] feel in the matter'.

Demonstration, extension, validation and education were four key functions of a horticultural show, and they enabled the first generation of European settlers to claim the high ground. Success as an exhibitor was a badge of modernity, and settlers could feel that they were citizens of a globalised society and economy. Horticultural societies organised competitions that attracted entries from town and country people, professional gardeners and amateurs, working people and businessmen, as well as the residents of small holdings and holders of large estates. At the very least, they were effective ways to publicise individual settlers' achievements and popular occasions to celebrate the successes of settler society in the new land. As early as 1848 the *Nelson Examiner* in its 29 February edition reported displays of fruit at a horticultural exhibition in a settlement not yet five years old as 'quite equal, we believe, to anything which could be produced with the most assiduous attention in any gardens in England'. Twenty years later, on 3 May 1867, *The Press* reported that at a formal dinner held in Christchurch the after-dinner speaker had 'complimented the gardeners of the Province on their great endeavours to excel the gardeners of the other provinces of New Zealand'.

Within a decade of the start of organised settlement in southern New Zealand, there were regular horticultural exhibitions in Christchurch, Lyttelton / Port Victoria, Akaroa, Timaru, Oamaru, Dunedin and Invercargill. In the early years, vegetables and fruit were the main exhibits, and a recent study of home gardening in Christchurch found that entries of decorative plants were more numerous once there were reliable supplies of locally grown food.[5] Committee decisions, venues and opening times for forthcoming exhibitions, as well as lists of prize winners, were reported in local newspapers and detailed summaries were reprinted by newspapers published in other population centres. Their competitive nature aroused keen public interest, and the prizes were spurs to greater effort by amateur and professional gardeners alike.

As well as annual and seasonal exhibitions, horticultural societies held committee meetings where members could view recently imported decorative plants and exchange information about their growth and propagation. Joseph Munnings, the owner of a carting business and general store in Christchurch, went to several of the local horticultural society's meetings and exhibitions, and wrote about them in his diary:[6] 'To Kohler's Gardens to see the flower show, very fair for the Province, good attendance' (27 December 1864); 'Boxing Day and Horticultural Show day, Friend Cole and myself looked in, in the evening for a short time, vegetables superior to the former show, flowers not so' (27 December 1865); and '3rd Horticultural Show of the season, very good show of both fruits and flowers' (28 February 1866). The following diary entry, dated 23 November 1865, was unusually full for Munnings and even more informative:

> Horticultural Show at the Drill Shed. Went in the evening and quite enjoyed it, for there I saw many things I would not have thought 'Canter' could boast of. All down the middle were high stools and tables covered with flowering shrubs and plants in full flower, around the side [were] fancy shrubs in pots, floral devices, bouquets, pansies, blossoms, vegetables of all descriptions, new potatoes kidney and round, Christchurch carried the palm for them, receiving first prize for each, so the bays [of Banks Peninsula] stood second. Some very fine rhubarb and cucumbers, broccolis, lettuce, globe artichokes, & etc. also some fine strawberries, ripe.

The depth of feeling aroused by such events was evident in the columns of local newspapers. One letter, published in *The Press* on 13 September 1866, was written by 'An Old Exhibitor' who queried a judge's qualifications. Eleven days later he followed with a longer letter that concluded: 'The best exhibitors have been driven away from some cause or other. I will presume that the cause is the committee sitting and making prizes to suit themselves; indeed, at the present time, they are the chief and principal exhibitors.' Public, frequently trenchant, criticism didn't stop there. In Christchurch, 'A Learner' questioned the judging of pansies in a letter to *The Press* on 19 December 1866, asking if the judges were 'justified in refusing prizes to all flowers which do not form perfect circles', and wondering if 'a blotch on the lower petal alone . . . is or is not a disqualification'. He was not the only exhibitor to raise such *recherché* concerns. At least one newspaper correspondent complained that there were not three separate classes: one for professional gardeners, a second for gentry who employed gardeners and a third for cottage gardeners. Even in ploughing matches, which were held during the winter months across southern New Zealand, a professional ploughman's success was viewed as a disincentive to the worthy amateur.[7]

LIVESTOCK SALES AND AGRICULTURAL AND PASTORAL SHOWS

If they were to do well in the new land it was essential for members of settler families to glean as much information as they could from all available sources, to observe what was happening in the district and to ascertain what they might achieve on their own properties. Newspapers reprinted articles from national and international publications, occasionally comparing agricultural developments overseas with the situation in New Zealand. On 24 August 1867, *The Press* published an extended report by Professor John Williams about an agricultural exhibition held in Aarhus, Denmark, which he had recently attended in an official capacity. The article had previously appeared in the Australian newspaper, the *Monaro Mercury*, and it highlighted the well-developed state of dairy farming in Denmark.

New Zealand had its own budding livestock and pastoral shows, which provided a more direct way for settlers and farmers to form opinions about developments, breeding capacities, and quality of stock and crops. By attending scheduled sales of livestock in large and small provincial centres, rural people were able to compare their animals with those of their neighbours, which was an important activity even when they had no intention of buying replacement or selling surplus stock. As Cody senior wrote in the family diary on 4 March 1915, 'was at Winton Stock Sale today. Fair lambs sold at 14/6 to 16/7 [for] good, wethers from 15/- to 19/- for fair ones whilst 24/6 was paid for one line of 6 and 8 tooth wethers of very large size'.[8]

The regional shows mounted by agricultural and pastoral associations played a major role in this by fostering displays of home-made food, livestock and crop plants. As well as 'encouraging the good breeding of horses, cattle and all other kinds of [live]stock',[9] they encouraged manufacturers of locally designed goods. The Canterbury Agricultural and Pastoral Association was formally founded in 1863, and its predecessors were an exhibition of merino sheep held at Shepherd's Bush in 1859, a similar event in Ashburton the following year, then 'a show of enlarged character [that] took place in Mr Justice Gresson's paddock in Latimer Square [Christchurch]'. The first issue of the *New Zealand Country Journal* included a report on that show taken from the 25 October 1852 issue of the *Lyttelton Times*, in which the writer criticised the quality of the exhibited sheep but praised the horses on display.[10] The many agricultural and pastoral associations that had sprung up across the country during the nineteenth century played a significant role in the development of rural New Zealand. On 10 May 1864, *The Press* reprinted a report taken from the 26 April issue of *The Colonist* about the annual exhibition of the Nelson Agricultural Association, held in Richmond on 21 April. It noted fewer entries than in the previous year's show because so many local men had decamped to the Wakamarina gold diggings. It also commented critically on the inferior quality of some entries:

> With the exception of oats and barley, the display in the grain department did not say much for the enterprise of our farmers . . . . none of the rye grass was deemed worthy of a prize. The unusual wetness of the season, when the grass was about to flower, appears to have had a bad effect on the seed, the experience of growers being that the seed this year is unusually light . . . . The continuance of the blight renders a bag of turnips a sight for 'sair een' [sore eyes] and provokes reminiscences of the years when acres of them could be seen on the farms.

In 1865 *The Press* published an advertisement on 12 April for the forthcoming Canterbury Agricultural and Pastoral Association exhibition. It listed eight classes for root crops, seven for grains and pulses, six for grass and clover seed, five for poultry, three for hay and two for dairy produce, 'All specimens exhibited to be the produce of the Province.' An entry dated 1 December 1888 in James Preston's diary for Longlands Station in the Maniototo hinted at what was involved for a prospective exhibitor: 'Cleaned grass seed for the show.'

Other landholders wrote in their diaries about mustering their best sheep and cattle and spending several days preparing them for display.

On 7 March 1866, *The Press* re-published a piece from the *Otago Daily Times* about the first exhibition of the Agricultural and Pastoral Society of Otago, which was held in Dunedin on 1 and 2 March of that year. The report referred to an earlier show linked to the New Zealand Industrial Exhibition of 1865 and went on to say that the society had been formed to continue that work: 'The keenness with which the merits of the exhibits were discussed proved the deep interest which the general public, as well as station holders, farmers and breeders take in the "points" of the different breeds and their improvement.' The exhibition attracted nearly 80 entries of sheep, 60 of cattle, 55 of horses and 25–30 of pigs, as well as a 'well selected, comprehensive and good' collection of implements for use on farm and station, and was attended by between 1200 and 1500 people.

In a brief news item published on 9 March 1865, *The Press* reported that the trophies for prizes awarded at the 1863 and 1864 shows had recently arrived from England, and the editor expressed the hope that 'the public will support the [Agricultural and Pastoral] Association in its endeavours to carry out its important object of creating an annual competition, in exhibiting animals and machinery, in order to produce an improvement in the breed of the former and manufacture of the latter'. In August of that year the newspaper published the following summary of 'the special objects' of the Northern Agricultural and Pastoral Association:

(1) To hold meetings at different centres between the Waimakariri and Hurunui rivers and to encourage, by the distribution of prizes and other means, the best modes of farm cultivation and the improvement of livestock. (2) To correspond with similar associations in various parts of the world, and to embody the scientific and other useful information relative to agriculture obtained in this and other ways in a journal or report, to be published by the society and distributed amongst members. (3) To procure reliable analyses of soils and manures with a view to increasing the productive powers of the country. (4) To encourage the manufacture and cheap conveyance of materials for building and general farm purposes. (5) To encourage the application of science to the construction of implements for the various purposes of agriculture, and the establishment of manufactures which derive their raw material from the produce of farm or

sheep run. (6) To procure and disseminate information as to the best means of checking the destructive ravages of insects, and the spread of noxious weeds, and to guard against the introduction or spread of disease amongst live stock. (7) To encourage the planting of the most suitable kinds of forest trees for the improvement of the country. (8) To collect and record reliable statistics of agriculture.

To those ends, a provisional organising committee of 45 members was formed and the annual subscription fixed at one guinea.

These 'special objects' were manifest in an advertisement placed by the Northern Agricultural and Pastoral Association, published by *The Press* on 1 October 1867, regarding the exhibition to be held at Kaiapoi on 6 November. The catalogue of prizes covered 26 classes for sheep, 22 for horses, sixteen for pigeons and poultry, seven each for cattle and pigs, and two for dogs (one for sheep dogs and the other for cattle dogs). The prize list for implements, tools and useful products made or grown in Canterbury was equally impressive: sixteen classes for farm implements, fourteen for manufactured goods, seven for dairy produce and two for harnesses. Two weeks later the Northern Association decided that 'a prize should be given for the best kept and managed farm in the district . . . . [and] prizes might be offered for the general management of best farms as well as for the best cultivated crops of roots or grain'. The reported reason was that because such a scheme worked well 'in the home country [the British Isles], it would be found to work equally well in Canterbury'.

The frequency, scope and detail of reports published in newspapers indicate the depth of public interest in shows and exhibitions, and in its edition of 10 November 1867, *The Press* printed an account of the recent agricultural and pastoral association show attended by almost 4000 people and described by the newspaper as 'the most successful one that has ever been given by the association'. The newspaper noted, however, that 'only four sheep dogs [had been] brought forward to contest the award offered, which is to be wondered at, seeing that Canterbury is the pastoral province of New Zealand. We had hoped to see at least a score entered.' The report mentioned fewer implements on display than in previous shows, but praised the collection of commercial ploughs and harrows made in Kaiapoi, horse-drawn carts made in Christchurch and the diverse local and imported woods used in their construction, the technically advanced wool presses made by the Canterbury Foundry, and a locally made butter churn which, in the manufacturer's extravagant prose, allowed its user to 'read the newspaper, smoke their pipe, nurse the baby and churn the butter at the same time'. On 27 December 1867, *The Press* published a long letter from R. D. Bust in which he called for a greater involvement of the Canterbury Agricultural and Pastoral Association in seeking solutions to the practical problems faced by grain growers in the region.

In that spirit, on 20 October 1879, Charles Tripp of Orari Station wrote to the editor of the *Timaru Herald*:

> Timaru A & P Association
>
> Dear Sir, Twelve months since I drew the attention of the Society to the great benefit exhibitors would derive if the Society would have cards printed and attached to each pen with questions to be answered by the judges, such as: Merino sheep – What defects are noticeable in the sheep? For instance: condition, form, length of staple, quality, colour of yoke. If these questions are replied to not only exhibitors but others will derive benefit. Without it most of us may be on the wrong track for years.

The following year, on 1 September 1880, he informed the secretary of the Canterbury Agricultural and Pastoral Association in Christchurch that he was willing to donate £10 to the inventor of a gorse-cutting machine that could be drawn by one horse and cost not more than £50.

Wherever they were held in southern New Zealand, agricultural and pastoral association shows were directed by local people with a vision for the orderly development of their respective districts. In its 26 November 1866 edition, *The Press* reprinted a short notice from the *Timaru Herald* about a meeting four days earlier of the organising committee for the Timaru Agricultural and Pastoral Association. That body's liabilities amounted to £155 10s and its assets (mostly from subscriptions and life members' donations) were £198 9s, leaving a surplus 'of a little over £40 to meet expenses for prizes etc', but the president 'hoped that more members and subscribers would come forward, so as to enable the association better to carry out the objectives in view'.

Judging at the shows was a serious matter. In 1885 the *New Zealand Country Journal* reprinted a piece that had been drawn from an article first published in the *American Live Stock Journal* about procedures used in show-yard judging at an exhibition of the American Clydesdale Association recently held in Chicago. Four years later, in 1889, an anonymous article drew on a piece by Lord

Combermere, first published in *Baily's Magazine*, in which he laid out judging procedures for show horses. The editor introduced it to the *Journal*'s subscribers across southern New Zealand with these words: 'At this season of the year when shows are the order of the day this article should be read with pleasure and profit.'[11]

SOCIAL ACTIVITIES AND COMMUNITY SERVICE

The diverse outside economic and social activities of settler families during the second half of the nineteenth and the early years of the twentieth century are illustrated in Figure 7.1 (overleaf). Nine of the farm and station diaries that I read contained relevant detail, and covered spans of at least five years.[12] Although two successive managers' letter books and three annual diaries for Ida Valley Station in Central Otago have been preserved, the only reference they contained to participation in organised off-property activities was the attendance of one of the two managers at a property sale in western Southland. On 19 October 1865, while he was still employed at Otematata Station in the Waitaki Valley, John Wither wrote in his diary, 'exhibited merino sheep, but did not win a prize at the Oamaru Show'. He shifted to Sunnyside in the Lake Wakatipu basin in 1871 and from then until 1890 recorded attending livestock sales and land auctions across the district as well as hosting church picnics on his property. On one occasion he exhibited sheep, potatoes and butter at an unnamed agricultural and pastoral association show, for which he obtained three first prizes. For such a long, detailed and continuous record of daily activities on the land, this is a short list of off-property activities. Whatever reasons there might have been for that, they are unlikely to have involved access difficulties. Sunnyside was served by as many as three lake steamers – which were hired by Wither each autumn to ferry bales of scoured wool, sheep and rabbit skins from his property to the railhead at Kingston – and the family owned a small vessel for regular trips across Lake Wakatipu to Queenstown. The daily entries in Wither's diaries suggest a serious, Calvanistic man who valued self-sufficiency, cherished his family, and saw no particular virtue in competitions and close social contacts. He was, however, a caring and compassionate neighbour. On 14 September 1883 he recorded, 'Mr Hassell's baby died this morning. Went over and made a coffin for the body, and went over to [Queens]town

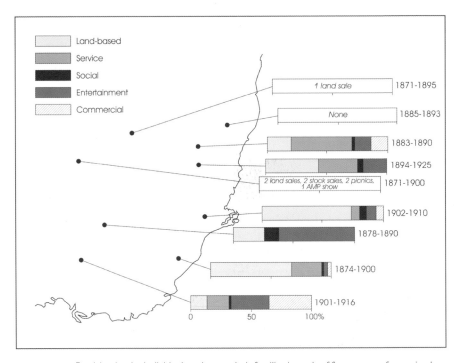

FIGURE 7.1 Participation by individual settlers or their families in each of five groups of organised activity, expressed as percentages of all days spent off-property in such activities over several years: (1) land-based activities, such as agricultural and pastoral association shows, horticultural shows and ploughing matches; (2) service on local body and charitable organisations; (3) participation in church, hospital and school committees; (4) organisation of or attendance at dances and socials, community picnics, horse races and sports days; and (5) attendance at auction sales and related commercial activities. In these settlers' diaries, references to church attendance are incomplete so this important social activity has not been included. SOURCE: SEE NOTE 12

with him by [the lake steamer] Antrim.' The next day he attended the inquest and burial. His wife, Marion, had spent the previous two days comforting the bereaved mother.

David Bryce, who farmed midway between Milton and Balclutha at the same time as Wither, left a comprehensive account of his farming, business and social activities between 1874 and 1900, when he served as butcher and baker, sold eggs and butter, provided basic veterinary services to neighbouring farmers, helped to manage the local school and church, and was closely involved with diverse rural organisations. One of them, the Mount Stewart Road Board, required his attendance at monthly meetings. In November 1899, Bryce was

Ploughman and team of horses competing at a ploughing match, place unknown
but probably northern Southland, late nineteenth or early twentieth century.
HOCKEN COLLECTIONS, UARE TAOKA O HĀKENA, UNIVERSITY OF OTAGO

nominated to represent the Crichton Riding on the local county council but
was not elected. For several years he served on the organising committee for the
Clutha and Matua Agricultural and Pastoral Association, which held its annual
main and 'walking' shows at Balclutha, and an entry in his diaries recorded
that he exhibited four horses, for which he received a first and a second prize,
in November 1873. In December 1888 he entered 'two lots of butter, two loaves,
scone and oat cake', and the following year won third prize for the best dry cow
or cow of any age on display. At the November 1899 show, his son exhibited a
selection of crossbreed sheep, winning one first, one second and one third prize
for them. At the same show, Bryce senior bought twenty prize-winning sheep
shown by John Begg. Most years, David Bryce, his brothers and sons travelled

by train to agricultural and pastoral association shows throughout the south-eastern South Island – in his diaries he mentioned exhibitions in Oamaru, Dunedin, Mosgiel, Milton, Waitahuna, Balclutha, Gore and Invercargill – and they usually exhibited livestock in shows close to the farm. He also attended ploughing matches at Inch Clutha and on the Tokomairiro Plains, and regularly attended property sales and sports days across the district. Few of Bryce's contemporaries left such a diverse record of participation in public life.

Joseph Davidson, who operated a small mixed cropping and livestock property near Waikaia, shared Bryce's interest in improving the local church and school. Between 1878 and 1900 he attended ploughing matches across Southland and occasionally went to agricultural and pastoral association shows in Invercargill, Gore and Riversdale. He was also closely involved with horse racing at Waikaia, Gore and Riversdale, and either attended or helped organise sports days, community picnics, balls, and 'soirees' in the church hall.

James Preston kept a detailed record of his diverse off-property activities while he was living at Longlands Station in the Maniototo between 1883 and 1890. He was an elected member of the county council; frequently exhibited in, served on the committee of, and attended agricultural and pastoral association shows in Naseby as well as those in more distant centres; and was active in a wide range of church, local government, sporting and promotional activities, including a visit to Dunedin to select books for the lending library then being established at Naseby.

The McMaster family, who farmed near Tokarahi in the Waitaki Valley, did not refer to off-property activities in their diaries between 1885 and 1893, whereas the Ramsay family at Hyde, on the fringes of Central Otago, whose diaries cover the period 1894–1925, were active in farm discussion groups, a medical club set up to cover its members' basic health costs, school picnics, sports days and local body committees. The Cody family at Lime Hills in northern Southland kept records of their off-property activities between 1901 and 1916, and they showed a broadly similar mix to those of the other individuals and families represented in Figure 7.1, although with relatively greater participation in and attendance at auctions and scheduled sales of horses and sheep. They successfully exhibited potatoes, mangolds, turnips and other farm produce at regional agricultural and pastoral association shows, participated in a wide range of church, sports, local body and social organisations close to their home, and went to auctions when neighbours' household effects and farm

Family picnic at Mount Horrible, south Canterbury, *circa* 1900.
6992, SOUTH CANTERBURY MUSEUM, TIMARU

implements were put up for sale. When he was a young man, Dickson Jardine worked on farms and pastoral properties thoughout the Waitaki Valley and in his diaries described social visits to nearby towns, including planned participation in sports days in Kurow (26 October 1902) and Oamaru (26 December 1903) that had to be cancelled or cut short on account of rain. He did not record attendance at church services or involvement in organising community functions.[13] The ninth of those settlers, James Miller, whose farm was on the northern flanks of Maungatua in east Otago, was active in a wide range of organised activities in the Outram area, and between 1902 and 1910 he and his family participated in committee meetings and attended several agricultural and pastoral association shows, flower shows and horticultural society meetings, and participated in school committees, local body agencies and community service.

Organised entertainments such as horse racing, rugby matches, school and church picnics, dances and soirées, school concerts and sports days, provided welcome opportunities for settler families and their employees to meet and share information, experience and knowledge. Some settlers were active in lodges, volunteer service organisations, school and church committees, and others served terms as elected members of local bodies and boards responsible for rural roads, water races and drainage, the relativities of the different types of off-property activity varying between settler families. Once telegraphic communications had made possible the ready exchange of information between large and small regional centres, and reliable transport systems permitted timely deliveries of newspapers and magazines, settler families became more closely linked to regional, national and international streams of activity, discovery and debate.

LEARNING ABOUT THE GOOD AND EXPERIENCING THE BAD

During the last four decades of the nineteenth century, settlers in southern New Zealand were kept well informed about events elsewhere in New Zealand, as well as in Australia, Great Britain, Western Europe and North America. Cargo and passenger ships brought mail, newspapers, books and other information from their places of birth or from where dispersed family members and close relatives were living, and the national telegraphic system and undersea cables enabled local newspapers to carry timely news of environmental events elsewhere in the country, important political decisions, prices for primary products on the London market, and information about what farmers and station holders were doing on the land elsewhere in New Zealand and in other parts of the world. The lineaments of a globalised economy were in place, and the residents of southern New Zealand were able to participate in it. Developments in science and engineering progressively eased the settler's task yet, paradoxically, blinded them to the real challenge: how to farm their properties in an ecologically appropriate and environmentally sustainable manner. For that, they needed to experience for themselves the environmental hazards of southern New Zealand and discover the vulnerabilities of their properties. Agricultural and pastoral displays and horticultural exhibitions showed them what they could achieve when everything was working to their advantage, but it was their learning on the job,[14] supported by what they had heard and read, that drove them to refine their operations to rise to the challenges of spells of unusually severe weather.

After tussock and shrublands had been cleared, wetlands drained, forest remnants reduced, fences and hedges established, and arable land ploughed, rural areas in the lowlands and low hill country of southern New Zealand might have struck settlers as simple ecological systems, but residents remained subject to spells of unusual weather, nutrient depletion and soil erosion. Land that had once supported relatively large flocks of sheep was unable to carry as many animals, some introduced species were beginning to behave in unexpected ways, and landholders regularly faced new challenges to the economies of their properties, including periods of poor market prices for meat and wool. They had to experience the good with the bad, and learn from their observations as well as those of their neighbours if they and their families were to thrive in the new land. This process could not be rushed.

CHAPTER EIGHT

Rural People
Continuing to Learn
about their Environments

After relocating to a new place, or when experiencing accelerated environmental change, living things encounter new opportunities and challenges and must respond effectively to them if they are to survive. During the second half of the nineteenth century the publications of Charles Darwin influenced how the biological sciences dealt with this, but applications of his evolutionary thinking to human societies proved less successful. Early in the twentieth century the American anthropo-geographer Ellen Churchill Semple reacted to the intellectual strait-jacket of 'environmental determinism' and proposed the much less doctrinaire notion of 'environmental influences' acting on people. To Semple, 'the earth whispers solutions to our environmental problems'.[1] Six decades later, two North American geographers, Yi Fu Tuan and Edward Relph, proposed that people invest time, money and effort into creating places full of meaning and significance to themselves, their families, contemporaries and successors,[2] making them the principal agents of environmental change. These investments mean that present and future residents can enjoy the sense of being psychologically attached to a particular place. Relph argued that people create landscapes of humanised places out of culturally neutral spaces by, amongst many such actions, building roads and houses, planting trees and shrubs, setting aside areas for recreation and exploration, passing on traditions and legends, and ensuring diverse opportunities for future generations to live and work happily there. In the words of the British geographer, Ronald Johnson, 'people make landscapes and landscapes make people'.[3] Relph distinguished between authentic and inauthentic place-making. The former involves the activities, legends and story-telling of successive generations of residents, enabling them and their progeny to feel native-rooted in a particular part of the world. In contrast, inauthentic place-making is often the consequence of organisations and powerful individuals deciding to erect structures to commemorate historical events without first involving residents. The first of these is evident in how successive generations of station families in the South Island high country have come to view and occupy their geographically isolated properties, a process that the American social anthropologist Michelle Dominy elegantly documented in her book about such a property in the mountains of Canterbury.[4]

Environmental changes trigger human responses, and throughout this book I have taken the stance that a landscape, farm or station can be viewed as a system: an organised suite of living things as well as pools of and channels for physical resources. The nineteenth-century German plant physiologist, Justus Liebig, investigated the mineral nutrition of plants, and his work soon became known around the world.[5] Its significance for New Zealand was evident to James Hector, who calculated that in 1891 the mineral nutrient content of exported meat weighed about one million pounds,[6] much more than was normally released during a year by nitrogen fixation and rock weathering in and immediately below the soil layer. Had he computed the average amounts of water needed to produce a side of mutton, a bale of wool or a bushel of grain, he would have been even more perturbed by the magnitude of environmental subsidies for the nation's exports.

During the 1930s the leading British ecologist Arthur Tansley popularised the word 'ecosystem' for the interactive web of living things in an area and the environmental resources that sustain them.[7] The American geographer Harlan Barrows implied much the same in his argument for geography as 'human ecology',[8] as did James Lovelock when he named the earth system after the Greek goddess, Gaia, and called for a new approach to the earth and atmospheric sciences. Lovelock was especially interested in diagnosing and treating the earth's environmental problems, most of which have been directly or indirectly caused by people.[9] Ecology developed rapidly during the twentieth century and five of its core concepts remain important. The first of these concerns the close functional relationships between a living thing and its physical environment. The second is that the diverse living things of a mature ecological system cohere in recognisable ways – associations, communities, ecosystems and biomes – and are not simply random aggregates. The third stems from the field research of the British ecologist Charles Elton who proposed that each natural ecosystem can be represented by a pyramid assembled from distinctive layers: green plants, herbivores, carnivores and decomposers. Through his field work in the Canadian Arctic and Great Britain, he found that each of those layers is characterised by the number of individuals of all species present and the total biomass, which allowed him to represent an ecosystem by a pyramid of numbers or a pyramid of biomass, and to use this construction to compare ecosystems in different parts of the world. Fourthly, after an episode of environmental disturbance – whether caused naturally or brought about by people, their plants and animals – the ensemble of plant and animal species will spontaneously change, and this process will last until a stable system has developed. Plant ecologists term this process 'succession', and it happens wherever land plants grow. While we may be able to follow or infer the progress of successional change, it has proven difficult to predict which species will be involved or even how long the process will take. Fifthly, ecology encourages us to think holistically. Even if we touch only one element or modify just one function of an ecosystem, then our actions will have a further impact elsewhere in the system.

The roots of a closely allied discipline, plant geography, penetrate even deeper into our past, and from the eighteenth century onwards interest in this subject was spurred by gardening, fostered by botanical exploration in distant lands and enhanced by the scholarly publications of staff at the Berlin, Geneva, Kew and Paris botanical gardens. Strabo, a Greco-Roman scholar active in the last two decades BC and the first two AD, had described the vegetation cover of places around the Mediterranean basin,[10] but the foundations for the scientific study of plant distributions were laid in the late eighteenth century by Carl Linnaeus and his students, and a short time later by Alexander von Humboldt.[11] Charles Darwin knew about their findings and corresponded with such leading figures as Augustin and Alphonse de Candolle in Geneva, Joseph Hooker at

Kew Gardens on the outskirts of London, and Asa Gray at Harvard University in the United States of America.

A core principal of biogeography is that widely separated parts of the world are occupied by different ensembles of plant and animal species, and this was certainly evident to the British-born naturalist Joseph Banks when he visited New Zealand as a member of Captain James Cook's first expedition to the southern hemisphere.[12] He encountered a large archipelago that had long been geographically isolated from potential sources of plants and animals by broad stretches of ocean. That situation ended with the arrival of the first Polynesian people a millennium ago and, starting in the late eighteenth century, when people of European ancestry began to settle the land, bringing with them many species of living things. A little later, Darwin expressed concern that the indigenous biota would inevitably succumb to this tide of novel plant and animal species because the long period of geographical isolation had, he believed, increased the vulnerability of native New Zealand species to population decline, possibly even extinction, in the face of biologically superior newcomers. Although awareness of the findings of biogeographers increased amongst biologists and gardening enthusiasts in New Zealand as elsewhere, relatively little of that knowledge appears to have filtered down to pioneer farmers and station holders.

THE PIONEER PROPERTY AS A SYSTEM

The landscapes of southern New Zealand were mosaics of extensive and small-area environmental systems when the European settlers arrived. Some of these systems were not greatly affected as settlers moved in and began the process of environmental transformation; others ended up at different stages along the path to becoming productive economic units; and yet more soon became places where environmental transformation was virtually complete. The processes of transformation employed by pioneer landholders meant that on any farm or station, some environmental features were erased, some were in a state of flux, and the balance had given way to novel systems of plants and animals. Pioneer landholders diverted streams, drained marshy depressions, levelled surface irregularities and planted surveyed rows or clusters of trees and shrubs to modify atmospheric conditions near the ground and to ensure a more

Haldon Station homestead, hired men's quarters and out-buildings, 1868–78, with recently burned tussocks on the hill behind the housing. SOUTH CANTERBURY MUSEUM, TIMARU, 1888

congenial environment for people, garden and orchard plants, pastures, crops and livestock. Despite the scale and intensity of these activities, the principal topographic features remained largely unchanged. The greatest changes were to the topsoil, the climate near the ground, the hydrological regime of rivers and streams, and the geographical distribution of native plants and animals.

The dynamic environmental systems of a pioneer farm or station comprised crop and pasture plants as well as grazing animals, supported by stores of water and plant growth nutrients and inflows of water and solar radiation. Reservoirs and channels were key functional units in the environmental system of a pioneer property, and they mediated movements of energy, water and other material resources into and through it as well as beyond its borders. The soil layer was the primary reservoir for water and exchangeable plant nutrients, the main channels for which were across and below the soil surface and thence into living tissues or transported away in large and small rivers.

Although it, too, was driven by shortwave solar radiation in the visible part of the spectrum, the environmental system of a pioneer farm differed substantially from the natural system(s) it replaced, notably with more biomass – in the form of meat, hides and wool, as well as grain and other plant tissues – and nutrients leaving the area for consumption elsewhere in the country or overseas. Trade in primary products was not environmentally neutral, but carried a cost in the currencies of the ecological resources of energy, water and nutrient ions. By the final decade of the nineteenth century, most lowland farms in southern New Zealand were dependent on supplements from external sources of nitrogen, phosphorus, potassium, calcium and other essential plant nutrients to remain economically viable. The energy cost might have been recoverable within a few weeks of harvest, and replacement of the water component would have taken from days to months, but access to new nutrient ions to make up for those lost through trade would have taken as long as was needed for minerals in the area's rocks to weather chemically and for nutrient ions to become available to green plants. Pioneer farms depended on a small number of plant species that could produce large amounts of palatable and nutritious tissues for as long as possible during the year, and this required inherently fertile soils. Most of the pasture plants that the first generation of European settlers introduced did best where water and nutrients were not in short supply or could be supplemented by irrigation and top-dressing with organic manure, guano and mineral fertilisers.

In an ecological system, negative feedback restrains flows of energy and materials and positive feedback enhances them. Furthermore, negative feedback tends to stabilise the system by moderating its responses to external forces and inputs. In temperate areas where a thin layer of fine sediment rests on hard rock, if evaporation is neglected then plant growth will be primarily governed by the balance between rainfall and runoff. As the vegetation cover develops, the rate at which rain water flows off a slope and into a stream will decline because the developing soil and vegetation cover will progressively retain more water on site. The net effect of this will be to make more water available to green plants for longer after rain, thus enhancing their prospects of surviving a spell of below average rainfall.

The deleterious effects of enhanced positive feedback to a pioneer farm or station are evident in areas subject to frequent burning and heavy grazing.

Depending on their size, herds of cattle and flocks of sheep can deplete the vegetation cover, trample the topsoil and expose the area to loss of fine sediment after moderate to heavy rain. Heavily browsed plants may die when declines in the amount of fine soil particles and dead organic matter reduce the potential for soil water storage. For as long as frequent burning and heavy grazing continue, relatively more precipitation will evaporate or run off the surface and into streams, carrying decayed organic matter and fine sediment with it. In time this will lead to further reductions in biomass and soil materials.

Any ensemble of plants and animals risks being set back to an earlier developmental stage by fire, disease, flood and erosion, and the specifics vary from place to place. Pioneer landholders lessened the environmental risks to their properties by damming small streams for water supplies, clearing stream and river channels to facilitate discharge of flood waters, planting hedges and trees for shelter, controlling stock numbers to safeguard pastoral land, growing sufficient animal feed during the frost-free season to set aside some for consumption during winter, using fire judiciously as a grazing management tool,[13] and ensuring safe areas on their properties for people and livestock. One environmental lesson that took settlers several decades to learn was that the economic benefits that flow from modifying the environment carry direct and indirect costs, so each landholder had to decide for himself if a particular development project merited the initial financial outlay over the long run. A new ecological system needs time to develop, form buffers that can stabilise it and become a substantially self-regulating entity. While they were establishing improved pastures for their livestock, few settlers took the long view of landscape transformation, and none of the diaries and letter books that I read contained explicit acknowledgement of the environmental price they might later pay for having failed to do so.

The transformational structures that settlers installed on their properties included hedges and shelter belts to protect people, plants and livestock from inclement weather; hedges and post-and-wire fences to control access by livestock to cultivated fields and pastures; areas set aside for seasonal grazing and haymaking; channels to distribute water to livestock and for irrigating pastures; drains to manage the amount of water stored in shallow water bodies and the topsoil; and trees planted alongside rivers and streams to regulate flow rates. Insofar as steep or other difficult terrain was concerned, few landholders saw merit in retiring it from grazing, encouraging reversion to native ecosystems, or planting it with timber and decorative trees.

Today, civil society requires social and environmental impact assessments before individuals can embark on large development projects. I did not find evidence of even the rudiments of such forward thinking in any of the nineteenth-century diaries and letter books that I read, although some landholders and managers, such as the Scottish manager of Ida Valley Station in Central Otago, had begun to sense that an adverse environmental event in only one part of the property could place the larger operation at risk.[14]

FROM COLONISED SPACES TO HUMANISED PLACES

The first two generations of European settlers on the plains and low hill country of southern New Zealand came to an expanse of grassy vegetation, peppered with wetlands, as well as large and small remnants of forest in well-watered valleys sheltered from strong winds. There were large as well as small tracts of native forest in the low hill country of Canterbury and Otago, and more extensive forested blocks in Southland. Much of the area had scant shelter from hot, dry northwest gales in spring and summer or from cold, wet southwest blasts in winter and early spring. For all its licence, Henry Sewell's description of the Canterbury Plains in 1852 as a 'howling wilderness' is an understandable response to the environmental conditions he encountered in the young settlement.[15]

In the plains and low hill country of southern New Zealand, geographical arrangements of hills, flat ground, depressions, rivers, lakes and ponds gave shape and structure to the landscape, and its physical form was elaborated by plants. On that grid, settlers demarcated fields and pastures with hedges and planted trees to provide shelter for people and livestock. They also modified or replaced indigenous tussock grass and shrub communities, drained wetlands, and sowed palatable herbs, grasses, root and grain crops for consumption on the property or for sale. In doing so, they were inadvertently creating ecological opportunities for introduced weedy plants and early successional native species to increase in number and occupy a larger area. Some of the plant and animal species that were deliberately brought into the country proved desirable additions in certain circumstances and at particular times, but unanticipated nuisances in others: gorse and broom, Yorkshire fog and yarrow, rabbits and red deer are six of many examples. The capacity of the New Zealand environment to

A successful pioneer family with their hired male and female labour, Team's farm, Otaio, south Canterbury, late nineteenth century. 2002-1026-00049, WAIMATE HISTORICAL SOCIETY AND MUSEUM

throw up surprise after surprise struck the first generation of European settlers just as it does us.

During the second half of the nineteenth century, settlers had begun to learn that the New Zealand environment is a mostly fine-grained mosaic of ecological patches, one that calls for close matching of pasture plants and livestock with prevailing physical conditions. The one-size-fits-all model of land development did not work during the colonial period, although it became more feasible later on with the advent of mechanised land preparation, extensive irrigation and widespread application of mineral fertilisers to correct nutrient shortfalls. Another early environmental lesson was the critically important role of extreme weather conditions in the seasonal round of activities of a farm or station. British experience was not a uniformly good guide to this, and settlers in southern New Zealand discovered the importance of casting their nets widely in the Old World as well as in recently colonised territories around the globe for examples of good practice in farm management and stock rearing to show them how best to respond to a geographically variable climate and occasionally adverse weather.

Settlers not only transformed the landscapes of their properties but were also engaged in an experiment, albeit one without controls, carefully specified treatments or trial runs.[16] They were informed about international practice in agronomy, horticulture and pasture management, but were creating a humanised landscape inspired as much by theological thinking as scientific principles, which led them to act in ways that, in retrospect, strike us as deleterious. At the time of their introduction, few settlers would have imagined that the European rabbit and Douglas fir could become environmental pests: the first within a decade of its introduction and the latter a century later. Amongst settlers' strengths were their flexibility in responding to novel environmental problems. That trait was especially evident in the last three decades of the nineteenth century when individual landholders, government officials and administrators sought ways to control the innumerable rabbits then plaguing southern New Zealand. Even so it took until the 1930s, when soil erosion had become too widespread and severe to neglect, for the nation to recognise the many risks it had been running.[17] Ideally, settler society should have screened imported plants and animals, and allowed entry only to those that were unlikely to spread spontaneously from where they were planted or released. This did not happen for more than a century, and New Zealand is now home to more naturalised than native species of higher plants.[18]

Did settlers recognise the links between their transformative actions and accelerated erosion, physiological drought and flooding? Partial as well as complete vegetation clearance depleted reserves of decaying organic matter in the topsoil, altered soil structures, and reduced amounts of water retained after rain and snowmelt, thereby weakening those ecological buffers that can come into play when current rainfall is too little to satisfy a plant's water requirements. During the second half of the nineteenth century, southern New Zealand experienced extended episodes of meteorological drought, but entries in farm and station diaries and letter books did not allow me to distinguish between the effects of spells of below-average rainfall on the one hand and loss of water storage sites in the topsoil on the other. For that, I would have needed reliable measurements of precipitation, water loss by evaporation and transpiration, storage in the soil and through-flow. Even records of precipitation received and water levels in wells sunk for household supply would have been useful, but out of the farm and station diaries and letter books that I read only those kept by Edward Chudleigh, the Cody family and the two Scottish managers of Ida Valley Station mentioned the latter.

Nor did I find documentary evidence that settlers were aware of the paradox in their dependency on native plants and animals at a time when they were busily eradicating them. Despite the advocacy of such well-placed individuals as Thomas Potts of mid-Canterbury,[19] entries in settlers' diaries and letter books implied that few amongst them envisaged native species having a permanent place in the humanised landscapes then under development in southern New Zealand.

Relph's notion of place-making is implicit in Max Nicholson's description of the southern New Zealand lowlands during the 1970s as 'a countryside in search of a landscape', to which he added these challenging words, 'and no doubt will find a worthy one'.[20] Settlers' achievements in learning about the environments of southern New Zealand were steps along the path towards the creation of an economically viable and congenial cultural landscape for themselves and their families. Technical education and opportunities to share good practice were essential in this, and on 14 October 1867 the *Otago Daily Times* published a short piece by its Tokomairiro correspondent about the likely benefits of a chamber of agriculture, allied with the farmers' clubs then in operation, 'to consolidate as it were the scattered intelligence throughout the southern portion of Otago in one focus for really practical purposes'. This call on farmers to share good practice drew a favourable response from the Otago Provincial Secretary and Treasurer, whose letter to agricultural and pastoral associations across Otago was printed by the same newspaper on 4 December 1867. Officials were invited to comment in writing on 'the present condition of the agricultural interest and the means by which encouragement and assistance may be offered to it'. The Board of Agriculture in the Australian state of Victoria was proposed as a model for the province of Otago, and respondents were asked to indicate their support for such a body, summarise their views on the establishment of a model farm, then identify new farm products, likely processors of farm products and the most important features of an Otago beet sugar industry. Even at this early date there was clear recognition of the importance of a scientific approach to agriculture and pastoral farming, and in its 29 January 1868 edition the *Bruce Herald* reprinted an article taken from the *Pall Mall Gazette* about science and farming in Germany since the early 1830s. It included these key words: 'The great secret of the success of Prussian agriculture is diffused education and technical instruction.' In southern New Zealand, the 'Prussian model' was seen as a way to enhance the management skills and practical education of people on the land.

LEARNING IN THE SCHOOL OF HARD KNOCKS

Settlers progressively learned about the environments of their farms or stations by observing, reading, asking questions and listening, and the course of their learning is shown in Figure 8.1. That representational model shows how settlers learned about local, regional and national environments by making observations,

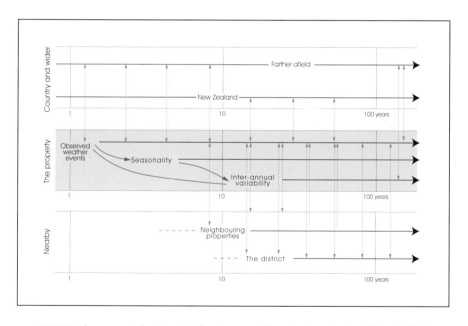

FIGURE 8.1 A representational model of environmental learning in early colonial times. Initially, rural settlers observed environmental conditions on their own properties (the shaded band) and compared them with what they had experienced elsewhere (the bands above and below the shaded band). As their awareness of seasonal effects grew stronger, they began to compare their experiences with those of people elsewhere in the district or farther away. Within a decade, they were more interested in discerning long-term variations on their properties and in the district. In time, rural people became adept at forecasting adverse weather events and drawing on that skill to manage their properties. SOURCE: SEE NOTE 21

discerning environmental signals, seeking and explaining patterns, calibrating particular environmental events against their own experiences as well as those of other people, and attempting to foretell weather conditions. The many geographical and historical comparisons recorded in farm and station diaries and other documentary sources justify the model.[21]

Entries in settlers' diaries also indicate more rapid discharge in streams and rivers after heavy rain and snowmelt towards the end of the nineteenth century than had been the case two or three decades earlier, which we can presume was a consequence of large-scale depletion of the vegetation cover, including conversion from geographically heterogeneous mixed tussock grass and shrubland ecosystems to extensive pastures of introduced grasses and broadleaf herbs. In the 1880s, Joseph Davidson of northern Southland started recording unusual flood events in his diaries:

> Came on to rain very heavy after dinner. Dome Creek rose very sudden – Creek falling [since then] as sudden as it rose. (14 May 1883)
> The [Dome] Creek still keeping very high. The stooks [of harvested grain] in the field has [sic] got a heavy drenching. If it [the weather] does not clear up soon it [Dome Creek] will get greatly discoloured. (25 February 1894)
> There was a very heavy hail storm passed over the Cattle Flat, hail stones very large – had the small gullys running full of water in a short time. (22 December 1898)
> Very bright and fine in forenoon – towards afternoon there was a heavy thunderstorm passed over – with heavy rain and hail. We seemed to escape the heaviest of it. It came down very heavy up Winding Creek and put it in flood. (15 January 1899)

Another indication of accelerated runoff after heavy rain or snowmelt is in a letter written by the manager of Ida Valley Station to the absentee landholder on 3 September 1895: 'In the last week of August we have had high winds and floods: most of the snow is gone and there are signs of grass [growth] . . . . Luckily the snow went away with wind and not with wet, and lately (except for sharp frosts at night) the weather has been mild . . . . During the late flood the creeks were higher than I have seen them, the Dovedale especially, and if it had not been for the heightening of the dam the flood must have gone right over it.'

Those experiences accord with observations of the McMaster family in the lower Waitaki Valley, where strong northwest winds followed by a southerly change early in the morning of 4 February 1894 brought persistent showers that turned to heavy rain that evening and resulted in flooded creeks by the following morning. Three weeks later, after 24 hours of very heavy rain, flooding recurred. Light showers during the evening of 11 November 1899, followed by a day of humid weather and an evening thunder storm, caused overnight flooding. McMaster clearly understood that a sharp burst of heavy rain may be shed from, and persistent light rain infiltrate, bare dry soil: 'Lots of rain last night, and raining off and on all today, nice quiet penetrating [rain]; best [there] has been for years' (11 February 1899). On 16 June 1887 a member of the McMaster family recorded, 'A regular flood. Rain coming down in torrents. Everything and everywhere getting flooded. River rising rapidly'; and six years later, in September 1893, it was this sequence of adverse weather:

Day fine; raining during last night. (10th)

Weather blowy. Morning warm and dry. (11th)

Weather wet .... Teams [of horses] idle; no work [done]. Fine evening. (12th)

Day fine, raining during night. (13th)

Raining all night from SE and NE. Too much rain. Awamoko very high. Heavy rain in evening. (14th)

Weather still disgraceful, NE and SE rains. Teams all idle. Awamoko in flood. (15th)

Much the same is evident in the Cody family diaries for the Lime Hills area of central Southland.[22] Their property stretched from low-lying riverine swamps, over well-drained downs to low tussock-covered hills. The former had been almost completely drained, and the latter virtually cleared of tussock and low shrubs, for cultivation by the early twentieth century. In their diaries, family members referred more frequently to flooding in 1911, 1912 and 1913 than in the previous ten years, probably a consequence of the loss of wetlands that would have stored water and released it steadily into rivers and streams over periods of days or weeks rather than almost immediately after heavy rain, as described in this entry on 28 March 1913: 'Raining heavily all day. Pa went out for Tom and Dave but could not cross river bridge for water .... Big flood, lane washed away, etc.' Much the same happened three years later when, on 7 September 1916, the family had to take a different route home than the one they had followed that morning because the river had risen sharply in the interim.

Pioneers and their immediate successors were understandably interested in quick-fix solutions. Hedges of gorse, hawthorn and broom were effective in areas where wood for fence posts and rails was expensive. They were functional within a decade, cheap to establish, confined livestock, and provided shelter for animals in times of adverse weather. Within a few years of planting, however, several of those virtues had become liabilities. Rapid shoot growth meant that a gorse or broom hedge had to be trimmed at least annually, and ideally immediately after flowering because viable seeds explosively expelled from ripe fruits allowed gorse to become an environmental weed of pastures, waste ground, roadsides and riverbeds.[23]

A rise in conservationist thinking throughout New Zealand led to significant changes in how settlers perceived and exploited the humanised landscapes of the south. In 1910, when James Preston's Goodwood property in east Otago was subdivided into smaller units, intending leaseholders were informed that at the end of the lease, they had to ensure at least 75 per cent of the arable land had been 'laid down with good healthy grasses, in sowing which the following quantities and varieties of good clean [weed-free] grass shall have been used to each acre, namely, fifteen pounds of ryegrass from old pasture, eight of Italian ryegrass, seven of cocksfoot, five of timothy, three of white clover and one pound each of cow grass and alsike cover'. The remaining area had to be in green crop or clean fallow. In addition, the leaseholder was required to 'protect and keep all timber, native bush and other trees from injury or destruction, and will not during the said term [twelve years from 8 October 1910] lop, cut down or remove any tree or shrub whatever on the said premises without the previous consent in writing of the landlord, AND ALSO will throughout the said term destroy and keep down noxious weeds and rabbits on the said premises and in all respects comply with the provisions of the Acts [of Parliament] for the time being in force for the destruction of noxious weeds and rabbits'. Finally, leaseholders were required to stack manure, straw and chaff for consumption on the property and were not permitted to plant new gorse hedges. Comparably rigorous requirements were in place elsewhere in southern New Zealand, as this tenancy agreement for Crown land on the Albury Settlement in south Canterbury shows:

> Plantations of trees [on the Albury Settlement, south Canterbury] go with the land .... The tenant is not authorised to cut down the plantations on his leasehold, except for thinning and for his own use, without the permission of the Commissioner of Crown Lands in writing, and it is a condition of the lease that trees equal in number to those cut shall be planted the following season.

And for the Kohika Settlement:

> Once a year throughout the term of the lease and at the proper season of the year, [the tenant shall] properly cut and trim all live fences thereon, and stub all gorse not growing as fences, and stub all broom and sweetbrier and other noxious plants.[24]

After three decades of relatively unregulated, even permissive, land management, settlers across southern New Zealand were subject to progressive control from government agencies: to manage hedges on road boundaries, maintain shelter plantings to reduce erosion in cultivated soils, control pest plants and animals, and protect remnants of native forest.

A slower and more considered approach to the propagation and selection of suitable hedge plants might have ensured fewer environmental problems for pioneer families and their successors, and something akin to a *SWOT – Strengths, Weaknesses, Opportunities and Threats* – analysis would have been helpful in this. Settlers had read about, even experienced, aspects of those four terms, but they were not presented by their financial and farming advisors[25] with a balanced account on which to base a rational decision. For more than a century, the rapid environmental transformation of rural New Zealand reflected the pace at which settlers made far-reaching decisions about the layout and economic activities of their properties, a process complicated by episodes of national economic depression curtailing opportunities for rational decision-making. Economic success required a mixture of forward thinking and painstaking management. Some settlers were able to ride out times of economic depression and could afford to develop their properties in ways that would diversify the landscape and conserve native species and their habitats. Others had no choice but to extract as large a surplus as they could from their land in order to survive. Despite this, they planted trees and shrubs in what had been an almost tree-less landscape, and even in the difficult times their newspapers and magazines advocated environmentally sensitive land use. That they achieved so much in just a few decades should be celebrated.

In practice, settlers had to experience the inherent environmental variability of southern New Zealand so that they could learn the rules of environmental stewardship for their respective areas. The widespread episodes of flooding, snowfall and drought that affected southern New Zealand during the second half of the nineteenth century could not have been foretold, and settlers had to discover for themselves the worst as well as the best environmental conditions of the new land, and learn how to manage their properties accordingly. For example, during a prolonged dry spell, John Wither deposited rubble and bush lawyer vines to armour the banks of a flood-prone stream near his homestead across Lake Wakatipu from Queenstown. A Scottish manager of Ida Valley Station was another. For several years he had observed the impact of below-average rainfall on the carrying capacity of the property's sunny faces, and wrote to the absentee landholder to propose a maximum flock size for the station, compatible with the periodically severe environmental conditions then experienced in that area, rather than the usual strategy of stocking to the maximum in good years and buying in replacement stock after a run of bad years had decimated numbers.[26] This environmentally protectionist spirit also underpinned the imposition of restrictions on shrub clearance and tree felling. For many rural settlers, the core lessons of conservation and environmental stewardship were learned in adversity.

Even before a national telegraphic network had been installed, settlers were alerted to local and international farming news items by the newspapers and magazines to which they subscribed. They also received letters, newspapers and magazines from family in the British Isles. Four groups of individuals stood out as advocates for rational landscape transformation: aware holders of large pastoral properties, such as John Hall of mid-Canterbury whose theatre of activity extended beyond the province;[27] committed scientists and administrators such as James Hector;[28] visionary newspaper editors such as those in charge of the *Canterbury Times, The Press*, the *Otago Witness* and the *Bruce Herald*; and knowledgeable importers of plants and seeds such as William Wilson in Christchurch. By the close of the nineteenth century, leadership was coming from local manufacturers of farm equipment, administrators such as Sir James Wilson,[29] scientists and educators such as Alfred Cockayne, and a cadre of government employees appointed to provide expert advice to landholders.

It is uncertain why early European settlers in southern New Zealand learned so little from Māori, and why so much of that knowledge concerned their immediate needs – chiefly where and how to cross rivers in flood, and the location of alpine passes – and not key features of the area's diverse landscapes. Environmental knowledge was managed differently in Pākehā and traditional Māori societies, and European settlers needed fluency in te reo Māori, the standing to receive traditional knowledge and a commitment to respect its ownership if they were to be granted access to it. Another important distinction was that European settlers were comfortable with numerical measurements, which they could add, subtract, multiply and divide then tabulate for drawing comparisons. In contrast, Māori environmental knowledge was orally transmitted and aphoristic, and it involved surrogates and proxies that experience over many generations had shown to be sensitive to particular environmental conditions. Their environmental knowledge came from close observation of the natural world, and they shared it with those who were eligible to receive it and authorised to transmit it. By contrast, the environmental knowledge of settler society came from routine observation and comparison. Sometimes it involved the use of calibrated rods (to measure, for example, the depth of flood waters), magnetic compasses (wind direction), thermometers (air temperature in sun or shade) and barometers (atmospheric pressure and trend), even crystals of rock salt to indicate unusually humid air, although early in the period of organised European settlement the environmental observations recorded in farm and station diaries were more usually couched in words than numbers. While both Māori and Pākehā were interested in changes between seasons and over runs of several years, when European settlers thought about trends they placed greater reliance on written records than recollections. It is tantalising to consider what the landscapes of southern New Zealand might have looked like had more Pākeha settlers developed the same sensitive relations with Māori as Edward Shortland did, and worked in a true partnership with iwi.

CHANGING RESPONSIBILITIES

Until the establishment of the Department of Agriculture in the late nineteenth century, settlers depended upon their own skills as observers, supplemented by expert advice published in newspapers and magazines. They were encouraged and occasionally chivied into taking certain actions by local newspaper editors who were fostering the spirit of continuous learning. Towards the end of the nineteenth century, the need for orderly development of export markets for New Zealand's primary products – initially wool and hides, then meat and dairy products, and later fruit – required farmers and station holders to produce standardised products in bulk for marketing purposes.[30] The task had become too great and too complicated to leave to individual farmers, as well as

207

the merchants, land agents and bankers who advised and supported them, and a cadre of trained scientific advisors hired by the state. Rural people were still encouraged by the newspapers and magazines that they read to observe, measure, test and experiment, but the call was not as strong as it had been a decade or more earlier. Even so, an editorial in the June 1907 issue of the *New Zealand Farmer, Stock and Station Journal* urged that 'every farmer should experiment for himself. He need not do it in a lavish or extravagant way, but on such a scale that he can prove for himself what answers on his farm, and he will then know what to go in for and what to avoid.' To those stirring words, the editor added that the cost was small, the satisfaction great and the knowledge acquired would amply repay any trouble.

In retrospect it would have been better had landholders and professional advisors worked much more closely with farmers, encouraging them to ensure their activities and management decisions were experimentally verified and scientifically justified, and had professional scientists been employed primarily to conduct technically demanding assays and experiments and to monitor long-term trends. In the February 1911 edition of the *New Zealand Farmer*, the editor referred to an address by the director of agriculture in South Australia to the agricultural section of the Science Congress that had recently been held in Sydney, where the speaker called for closer co-operation between farmers and professional scientists, and greater financial support for agricultural field stations. In many respects, James Wilson and Alfred Cockayne embodied this spirit in their work with the Department of Agriculture, where the latter was prominent in extension courses and farmer education. Cockayne's successor, Bruce Levy, sustained the educational activity but made a greater mark as an agronomist and a publicist for scientific agriculture. All three wrote about native plants as sources of nutritious herbage for sheep and cattle, but their prime concern was to advance pastoral farming in New Zealand rather than to promote native plants in the rural landscape. Commentators like J. B. and J. F. Armstrong and William Travers in Canterbury had earlier expressed concern over the parlous state of native species in the agricultural lowlands, but comprehensive strategies for involving native plants and animals in the rural landscapes of southern New Zealand would not come until well into the second half of the twentieth century.

If they even thought about the matter, for most European settlers the natural world was a place where desirable plants and animals, regardless of provenance, thrived. Many of the arrivals had come from long-settled parts of the British Isles with landscapes developed by people over several centuries. In New Zealand, native timber trees were a prime resource when settlers were establishing farms and sheep stations, yet were not perceived as long-term elements of the landscape that settlers wished to occupy. Some purposefully introduced plants failed settlers' expectations by behaving differently from populations of the same species in their home areas and becoming weedy. A smaller number did what was expected of them, posed few if any problems and were used by settlers instead of other potentially useful candidates. In the light of what we now know, the most regrettable aspects of landscape trans- formation during the first century of organised settlement were the shift from diversified to simplified plantings in pastures, shelter belts, orchards and decorative areas, and settlers' inability to recognise that a fine-textured physi- cal environment requires a comparably fine-grained land use. Settlers' great achievements, however, were their flexible responses to environmental prob- lems and the way they parcelled previously open rural landscapes with shrubs and trees. Even so, the latter benefit was diminished by settlers' economic reliance on a few species of fast-growing economic plants, inconsistent man- agement of them, and few procedures for replacing shelter and hedge plants as they reached maturity. There were ways to deal with this, but they involved knowledge of agronomy, conservation, ecology and forestry in a rural environ- mental setting.[31]

The words 'conservation' and 'nature' were rarely used in late nineteenth- century newspapers and farming magazines, and were absent from the farm and station diaries that I read. On the basis of what they wrote in their diaries and letter books, it appears that European settlers in southern New Zealand saw it as their God-given responsibility to create economically viable, func- tional, supportive and pleasing lowland landscapes with the plants and animals of their choice. Although roads, tracks and property boundaries were surveyed early in the period of organised settlement,[32] the first generation of rural settlers did not follow a fully specified landscape plan or conform to an official standard: that came with subdivision of the large estates, when prospective tenants were informed in writing what was expected of them. The outcome – a gridded array of productive fields enclosed by trees and shrubs, houses and out-buildings sur- rounded by decorative and productive food plants, criss-crossed by roads, and dotted with small settlements – was the precursor to the natural world as they believed it should be. Any other species in any other arrangements would have lacked significance in the overall scheme of things.

A century and a half ago, the first generation of European settlers set out to transform the landscapes of southern New Zealand, learning which plants and animals to use and how to undertake the process. Within a couple of generations they had achieved many of their personal goals, but few among them appreciated why contemporary commentators like J. B. and J. F. Armstrong advocated so vigorously for native grasses in productive pastures. By the 1890s settlers were making steady progress towards recognising and interpreting signals of adverse weather, selecting productive mixtures of pasture species appropriate to the different environmental conditions of their properties, choosing livestock varieties well suited to their land, deploying trees and shrubs to decorate their homesteads and provide shelter from adverse weather, and learning how to manage weedy plants and pest animals. It took a century for their successors to apply those lessons to a landscape in which native species might co-exist with introduced species, and where natural and semi-natural ecosystems could thrive alongside managed economically productive ecosystems (Table 8.1). In our time, the model of an ecologically functional landscape, where economic as well as conservational values are sustained, is increasingly evident but still not widespread. The processes and accumulated lessons of environmental learning are no less relevant today than they were a century and a half ago.

TABLE 8.1 Key features of system change on farms and stations in the tussock grass and shrubby lowlands of southern New Zealand.

System feature	Initial	Early colonial	Mid-20th century	In prospect
(−) feedback loops	strong	diminishing	weak	strengthening
(+) feedback loops	weak	increasing	strong	weakening
Buffering	strong	weakening	weak	strengthening
Responsiveness	slow	more rapid	rapid	slowing
Landscape mosaic	heterogeneous	less diverse	homogeneous	diversifying
Primary nutrient inputs	internal	some external	largely external	internal conservation values

When there is a stronger interest in re-establishing native plants and animals in the productive lowland landscapes of southern New Zealand, whereabouts will interested people obtain seeds, cuttings and rooted plants? That question is important because it is good international practice to sow or propagate local genotypes rather than to source them widely. After 150 years of burning, felling, ploughing and grazing, reserves of seed in the topsoil have been reduced, and there are few remaining areas of native or semi-native vegetation left in the lowlands from which native plants and animals might repopulate economically marginal land taken out of production. Wood pigeons and other native birds that dispersed tree seeds are now uncommon in the countryside, and interested people might need to assist the normal processes of succession by planting the tree, shrub and herbaceous species once found growing in the area to provide food and habitat for the more mobile native birds and insects. The alternative will be a mix of ecosystems unlike those previously found in the area: initially dominated by gorse and broom, then occupied by weedy adventives and pest animals as succession proceeds until mobile and less environmentally constrained native plant and animal species can become established. Once late successional systems dominated by native species, but including well-adapted introduced species, have developed in the midst of economically productive farms and pastoral properties across the rural landscapes of southern New Zealand, native animal and plant species should be able to follow habitat corridors or chains of habitat islands as they move between scattered refuge areas. It will be necessary for people to control the spread of several introduced plant and animal species and manage weedy plant and animal pests – a task that may never end – but a new generation of residents will learn the key lesson that sustainable and economically productive landscapes in the plains and low hill country of southern New Zealand should comprise functional, managed mosaics of mostly native and largely artificial ecological systems.

Words about Home Diaries and Letters, Commercial Transactions, Newspapers and Magazines

From the earliest years of organised settlement in southern New Zealand, rural people kept daily accounts of their life and work, as well as descriptions of environmental conditions in their home areas. Some diaries, like those of Edward Chudleigh and Edward Ward, were primarily for the information and entertainment of family in Great Britain.[1] Most, however, contained informative summaries of activities on the property and details of notable environmental, social and political events for future reference. Thus, Edward Ward recorded on 17 March 1851: 'Having read in the Australian papers of a hot wind at Port Philip [Victoria, Australia], which astonished inhabitants and raised the thermometer to ninety-two degrees on February 6th last [1851], I look back to my journal and find that we had the very same sort of day [in Christchurch], only that our thermometer rose to ninety-four degrees.' On 3 July 1889 the manager of Ida Valley Station in Central Otago wrote in a similar vein to the absentee landholder: 'The first winter I came here (1879) was by far the most severe I had experienced till this one, and my impression was that it was as severe. I find, however, on looking up the diary that it was not nearly so bad. We were ploughing until end of June [1879]; and in July, although the snow was deeper and fell frequently, it lay very little time and the longest spell of hard frost was seven days.'[2]

Farmers and pastoralists usually wrote their daily records on lined paper in bound notebooks, although some used school exercise books or pocket diaries with the seven days of a week set out on facing pages. By the late 1850s *pro forma* diaries were readily available from booksellers and rural supply companies. A few were published in England for use throughout Australasia, others were printed in Australia for use there and in New Zealand, and the rest were published in New Zealand. Most contained an introductory cluster of printed pages with information about matters of general interest: the names of living members of the royal family and colonial governors, phases of the moon, tides, excise on imported goods, important legislation, dates of public holidays and the like. In the more than 50 runs of farm and station diaries that I located (Figure A.1), there were eight *pro forma* diaries published by Letts in London or under licence by Cassell in Melbourne, six by Whitcombe & Tombs in Auckland or Christchurch, four by W. Collins in Auckland, one by J. Couchman in Wanganui, and two 'Scribbling Journals', one of which was printed by W. Collins and the other by an unnamed company.

Depending on page size, the writer had between three and ten centimetres of blank or ruled paper on which to record operational details, environmental observations and other information about the property. Despite its quirky spelling, inconsistent capitalisation and capricious punctuation, the following entry dated 15 October 1895 in the diaries mostly kept by David Bryce, who farmed a small property with his siblings and family at Lovells Flat midway between Milton and Balclutha in east Otago, indicates the diverse economic, environmental and social information that this documentary source often contains, and hints at the challenges facing the modern reader interested in extracting it: 'This was a very good day, was up around the sheep was thru the cut side sheep got 4 Old ewes lying dead skined them was down at the Station sent away a letter two Thomas M'Donald Stirling John was sowing grass seed on A.D. Hills in the back of the house was down at the Flat at Night at a Prayer Meeting in the School House Mr William Fraser wants a Heafer he ofered me

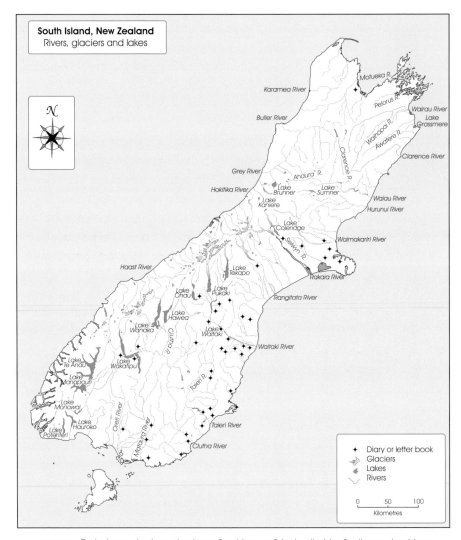

South Island, New Zealand
Rivers, glaciers and lakes

FIGURE A.1 Each dot marks the main place of residence of the landholder, family member(s) or employee(s) whose diaries and / or letter books were available to me.

£5 for young Leady David is harrowing at the Old Man's W. Patterson got 2 doz eggs & 1 lb butter.'[3]

The following entry from Bryce's diaries is reproduced verbatim (italics) then expanded, re-ordered and re-punctuated for ease of comprehension (normal font):

15 March 1899: this was a fine day was at home all day was down & got a ram & take
it up to Robert Wilson for the season s 10. put 2 ewes in the shed Paddock out of the
Windmill Paddock now & 5 horse we have got the Hastes ridge oats all in Braut 2
wethers on this side up the line very Bad with foot rot got 4 laves s1 & [?] lb tobako
laves s3 for the man got the account for the Flour £7.18.

15 March 1899: The weather was fine, and I was at home all day. I caught a ram
to take to Robert Wilson to service his breeding ewes for a fee of ten shillings
then moved two ewes from the Windmill Paddock to join five horses in the Shed
Paddock. We brought into the yard all the stooks of oats harvested at Hastie's
Ridge. Two of the wethers on the house side of the railway line have very bad foot
rot. I got four loaves of bread for one shilling and [one?] pound of leaf tobacco for
three shillings for the hired man. I received the account for flour purchases: £7
18s 0d.

Even after editing, farm and station diary entries can seem coded, but they were
for the information of the writer and his immediate family and there was manifestly no need to spell out every last detail.

Farm and station diaries were primarily aids to management, and what was
written depended on the writer and the occasion. Thus, on 1 December 1891,
James Preston of Haldon Station in the Mackenzie Country wrote: 'Fine day.
Started shearing with 8 shearers. J. Long & Tuby tablemen at 25/- per week.
W. Lavers, picker, 15/- per week. Jackson and Scoringe, pressers, 1/- per bale.
W. Blanchett, cook's mate, 20/- per week. H. James, belly [wool] picker, 15/-
per week. G. Green, W. Waddington and J. Murray, wool scour, 25/-.'[4] Some
entries, however, were more personal in tone, as in Preston's note written on
19 May 1899 about the lonely death of a rabbiter employed under contract on
Longlands Station in the Maniototo: 'All hands looking for Jimmy Ford. Found
his body in Houndburn Paddock, where it must have laid since 29th April.' A
draft of Preston's letter to F. W. Morrice at the Public Trust Office in Dunedin
has been preserved in his personal papers, and adds detail and poignancy to a
sorry tale:

Your letters of 25th and 29th inst. [regarding?] the property of James Ford,
deceased hand. At Constable Johnston's request, I sent my man to gather up
the [rabbit] traps and to bring anything of value to the [Longlands] Homestead.

The tent [in which Ford lived] was nearly rotten and the bedding was covered with filth. Ford's dog had been lying on the blanket for [indecipherable] weeks. Constable Johnston and I concluded to bury it as it stood, and Anderson, the man who bought the traps, said he would give me £1-0-0 for it unseen, so I took the offer. Constable Johnston told me that Mitchell would be the undertaker, and that burial expenses would be defrayed from the money left.[5]

Entries in farm and station diaries can also tell us about the writer's interests, lifestyle and aspirations. On 25 March 1912 a member of the Ramsay family, who farmed near Hyde on the fringes of Central Otago, wrote in the household diary, 'Mr MacGregor O[tago] E[ducation] B[oard], instructor in wool classing, gave a lecture in the school tonight. Jack, father and I attended. Father and I joined the class. A lecture is to be given weekly for ten weeks, fee £1.'[6] While most entries in the diaries available to me were laconic summaries of what was done on the property during the day and the people involved, they frequently contained brief accounts of the writer's environmental observations and impressions, reading and non-work activities; as well as visits by neighbours; the comings and goings of contract labour; when and how many sheep were shorn; shearers' and shed hands' names; numbers of bales of wool sent off the property; names and quantities of trees and shrubs brought in from nurseries for hedges and shelter belts; plants for kitchen garden and orchard; and decorative plants collected from nearby bushland, obtained from neighbours or bought from nurseries. Several rural people even inserted comments about national economic and political conditions, the state of the London market for wool, hides and rabbit skins, and auction prices for wool.

As records for future reference, virtually all the farm and station diaries that I read contained summaries of local and regional weather events as well as accounts of floods and droughts, with sufficient information about each for the writer or another resident to confirm recollections of earlier events, draw comparisons and identify recurrent patterns. In the digest of useful information printed on the opening pages of the *New Zealand Rough Diary No.3 1895*, printed by Whitcombe & Tombs in Christchurch, a short section on page 29 – which was reproduced largely without change in later issues of this publisher's *pro forma* diaries – included the following 'Plain Rules for Foretelling the Weather':

The Barometer Rises for southerly wind (including from SW by south to the eastward); for dry or less wet weather, for less wind, or for more than one of these changes. Except on a few occasions, when *rain* comes from the southward with *strong* wind.

A Thermometer Falls for change of wind towards *any* of the above directions.

The Barometer Falls for northerly wind (including from NE by the north to the westward); for wet weather in winter, for strong wind (in summer), or for more than one of these changes. Except on a few occasions, when *moderate* wind with rain comes from the southward.

A Thermometer Rises for a change of wind to the north.

Two decades later, the *Scribbling Diary for 1915*, published by Collins Brothers in Auckland, included weather forecasting guides such as these two from page 24: 'Change of wind may be expected when high upper clouds cover the sun, moon or stars in a direction different from that of lower clouds, or the wind felt below'; and 'Stormy weather may be expected when sea birds hang about the land, or fly inland.' The following weather guide was printed on page 9 in *Collins' Commercial Diary for 1923*:

The latitude of New Zealand in the Southern Hemisphere corresponds very closely with that of Italy in the Northern:

Auckland	Cape Passaro / Sicily
Wellington	Naples
Christchurch	Florence
Dunedin	Venice

Low pressure, with the barometer below 30 inches, usually brings more humid conditions; and while the barometer falls the wind is in the north and the weather is warm and wet. When the wind turns by the west to the south for the rise in the barometer, the weather is colder and sometimes very wet and snowy.

That reference to snow may be unique in the *pro forma* farm and station diaries published in this country. This published diary also repeats the previously common climatic comparison between New Zealand and the Mediterranean basin. None of the printed introductory materials in the commercially printed diaries available to me, however, referred to flood or drought, erosion or land instability.

Given the vital importance of the weather to people on the land, published guides such as these are not surprising. In their daily diary entries, settlers frequently distinguished between fog, mist, drizzle, steady rain, squally rain, hail and snow, and normally recorded the onset and duration of a prolonged wet period. Wind speed and direction, however, were less frequently noted, although cloud banks signalling the approach of a southerly change or a thunderstorm were reported. Charles Ayton, who lived in the Upper Taieri region of east Otago between 1899 and 1904, recorded twice-daily readings of barometric pressure.[7] Other settlers simply noted 'falling glass', 'glass very low', 'rising glass' or 'glass high'. What is curious about the poverty of actual measurements of atmospheric conditions is that intending settlers had been encouraged to bring with them mercury-in-glass barometers with in-built thermometers, and many had done so. Most references to temperature in diary entries were framed in terms of human comfort: for example, 'fearfully cold', 'bracing' and 'stiflingly hot'. Rural people usually recorded the depth and persistence of winter snow and described the environmental impact of flooding in nearby streams and rivers, but few of the earliest settlers referred to land instability consequential on vegetation clearance or cultivation of surface soils. It was not until the 1870s that they began regularly to report problems with erosion and declining soil fertility on their properties.

Eight runs of farm and station diaries and letter books are frequently cited in the chapters of this book, where each is identified by the name of the author and the property. Other details are in the endnotes or references.[8] The eight contained unusually detailed accounts of daily weather conditions made virtually without break over several years (Figure A.2). The earliest was written by Edward Chudleigh during his first four years in New Zealand (1862–65). It is a vigorous young man's diary written while he was employed as a stockman by the Acland family of Mount Peel Station in south Canterbury, and later posted to his mother in England. During his work-related travels throughout Canterbury, on the West Coast and in north Otago, Chudleigh recorded his observations of daily environmental conditions and the hazards posed by strong wind and heavy rain, drought and flood, summer heat and winter frost, and major falls of snow. John Wither's almost complete run of daily records for properties in north Otago, near Arrowtown and in the Lake Wakatipu basin began early in 1864 and extended into the twentieth century. The third set of daily manuscript records lasted almost without break from August 1866 to

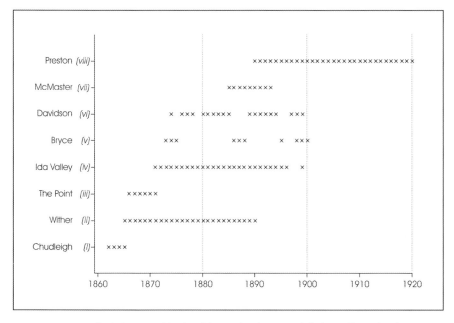

FIGURE A.2 Periods covered by the eight overlapping runs of diaries and letter books
that provided geographical and historical context: (i) Edward Chudleigh, (ii) John Wither,
(iii) The Point Station, (iv) Ida Valley Station, (v) David Bryce, (vi) Joseph Davidson,
(vii) the McMaster family, and (viii) James Preston and family. Coverage for complete
and part years is shown, and gaps signify missing volumes. SOURCE: SEE NOTE 8

May 1871, and was kept by an un-named resident of The Point Station in inland
mid-Canterbury. The letter books and diaries of two Scottish managers of Ida
Valley Station in Central Otago contained much useful environmental information for the years between 1871 and 1899. Albeit with several gaps, David
Bryce's diaries covered the period from October 1873 to December 1899 and
concerned the family farm that he and his brothers owned and operated at
Lovells Flat, midway between Milton and Balclutha in south Otago. Despite
gaps of weeks to years, the diaries kept primarily by Joseph Davidson covered
the years from 1874 to 1899 for his family's livestock and grain farm on the outskirts of Waikaia, formerly known as Switzers, in northern Southland. The set
of daily records kept by members of the McMaster family, who occupied a pastoral property at Tokarahi in inland north Otago, extended from 1885 to 1893.
The last of the eight sets of records was provided by the Preston family archives
for six pastoral properties in Southland, east Otago, the Maniototo, the upper

Waitaki Valley and the Mackenzie Country for various periods between 1890 and 1920.

Amongst the documentary sources for an inquiry into settlers' environmental knowledge, as well as how it grew and became more comprehensive with experience in the new land, settlers' business papers and personal letters can tell us much about environmental conditions, major weather events and floods, but they are the least readily available source of information. The dozen letter books that I traced were compilations of letters from station managers to absentee landholders, land agents in nearby large settlements, commercial suppliers of seeds and fertilisers, and newspaper publishers. Copies had been made with the aid of carbon paper or, more usually, by pressing the handwritten original on to a thin sheet of moistened tissue paper bound into a letter book. The impression left by indelible pencil or liquified ink can be read from the reverse side of the bound sheet, and legibility is often good.

Despite their limitations, considerable amounts of useful environmental information can be extracted from runs of nineteenth-century farm and station diaries and letter books. There is, however, an important *caveat*. What a resident wrote is only what he or she chose to record. What the individual might have observed or thought about an event, but did not set down in writing, remains unknown to us. Clearly, when using this material one must respect the injunction 'absence of evidence is not evidence of absence'.

Although relatively few seem to have been preserved, records of commercial transactions relating to a rural property proved informative and provided considerable insight into the economic cost and environmental consequences of the 'rabbit plagues' that affected southern New Zealand from the 1870s onwards. Detailed reports of floods, snowstorms and gales, as well as first-hand accounts of what residents had learned about the environments of their farm or station, were frequently published in local newspapers and reprinted by newspapers elsewhere in the country. Many newspapers retained correspondents in the goldfields of Central Otago and the West Coast. Their reports, along with lists of formal meteorological observations made at dispersed weather stations, were widely re-published for the information of readers across the South Island and beyond. Most newspapers had correspondence columns in which readers engaged in vigorous debates about matters of current interest, outlined their experiences on the land and requested advice about how to resolve practical problems. Regional newspapers often published extended extracts from ships' logs about weather conditions in coastal waters after major storms had lashed the country and surrounding seas, and reprinted relevant agricultural, environmental and scientific reports from newspapers and magazines published overseas. The accounts of daily life and work in the new land published by the *Canterbury Times* and the *Otago Witness* made these two newspapers important supplementary sources of information for most chapters.

From the 1880s onwards, numerous publications directed to rural people were produced, distributed and read in New Zealand.[9] One of them, the *New Zealand Country Journal*, was published between 1877 and 1898 by the Canterbury Agricultural and Pastoral Association, and it carried technical articles of special interest to well-informed landholders. The *New Zealand Farmer, Bee and Poultry Guide*, later renamed the *New Zealand Farmer*, was first published in Auckland in 1885, had subscribers throughout the colony and served a rural readership.[10] Both offered guidance and information to people on the land, fostered debate and promoted the sharing of good practice. The *New Zealand Journal of Agriculture* was later to become an important weekly forum for the exchange of useful information, observations and experience between the growing cadre of professional scientists and advisors and people on the land.

NOTES

Introduction

1   A. H. Clark, *The Invasion of New Zealand by People, Plants and Animals: The South Island*, Rutgers University Press, New Brunswick, 1949, p. 158.

2   K. B. Cumberland, 'A Century's Change: Natural to Cultural Vegetation in New Zealand', *Geographical Review*, vol. 31, 1941, pp. 529–54.

3   K. Anderson, *Predicting the Weather: Victorians and the Science of Meteorology*, University of Chicago Press, Chicago, 2005, contains sections about Robert FitzRoy's achievements; C. J. Burrows, *Julius Haast in the Southern Alps*, Canterbury University Press, Christchurch, 2005, discusses the life and work of Julius Haast; and S. Nathan and M. Varnham (eds), *The Amazing World of James Hector*, Awa Science, Wellington, 2008, is a collection of fifteen essays about James Hector's notable career. For an account of Charles Torlesse's early weather observations in mid-Canterbury, see P. Holland and B. Mooney, 'Wind and Water: Environmental Learning in Early Colonial New Zealand', *New Zealand Geographer*, vol. 62, 2006, pp. 39–49.

4   Anderson, *Predicting the Weather*, pp. 105–9.

5   Joseph Greenwood's diaries, southern North Island, Wellington, Banks Peninsula and north Canterbury (transcript): Alexander Turnbull Library, Wellington, MS-PAPERS-4882.

6   Joseph Munnings' diaries, Christchurch and environs (transcript): Canterbury Museum, Christchurch, ARC 1990.50, item 4.

7   L. J. Wild, *The Life and Times of Sir James Wilson of Bulls*, Whitcombe & Tombs, Christchurch, 1953.

8   The report that Munnings referred to in his diary had been published the previous July: W. T. Doyne, 'Report on the Rivers and Plains of Canterbury, New Zealand', *The Press*, 5 July 1864.

9   J. Hector, 'The Utility of Natural Science', *Otago Daily Times*, 24 October 1862.

10   James Preston's diaries, personal and business papers, Southland, Maniototo, east Otago, Waitaki Valley and Mackenzie Country: Hocken Collections, MS-1271 and MS-1272. This large collection of diaries, ledgers, letters and private papers was used by a descendant for a book about the Preston family in New Zealand: F. L. Preston, *A Family of Woolgatherers*, John McIndoe, Dunedin, 1978.

11   John Barnicoat's diaries, Nelson and eastern South Island (transcript): Hocken Collections, MS-1451.

12   Thomas Ferens' diaries, Waikouaiti, east Otago (transcript): Hocken Collections, MS-0440/016.

13   Francis Pillans' diaries, Inch Clutha, south Otago (transcript): Hocken Collections, MS-0616.

14   E. R. Ward, *The Journal of Edward Ward: 16 December 1850 to 22 June 1851*, Pegasus Press, Christchurch, 1951.

15   McMaster family diaries, letter books and cash ledgers, Tokarahi, Waitaki Valley, north Otago: Hocken Collections, MS-1011.

16   Ida Valley Station diaries and letter books, Central Otago: Hocken Collections, MS-0658.

17   'Hopeful', *'Taken in': Being a Sketch of New Zealand Life*, W. H. Allen, London, 1887.

18   Ward, *The Journal of Edward Ward*, p. 106.

19   J. W. Stack, *Through Canterbury and Otago with Bishop Harper in 1858–60*, Akaroa Mail, Akaroa, not dated, p. 3.

20   P. Holland, J. Williams and V. Wood, 'Learning About the Environment in Early Colonial New Zealand', in T. Brooking and E. Pawson (eds), *Seeds of Empire: The Environmental Transformation of New Zealand*, I. B. Tauris, London, 2011, pp. 34–50.

21   Except where formal names are given in the source, or are necessary for clarity, common names for plants and animals are used throughout the text. For lists of common and corresponding scientific names of plants, see E. R. Nicol (compiler), *Common Names of Plants of New Zealand*, Maanaki Whenua Press, Lincoln, 1997.

1   Māori Environmental Knowledge

1   D. N. T. King, J. Goff and A. Skinner, 'Maori Environmental Knowledge and Natural Hazards in Aotearoa-New Zealand', *Journal of the Royal Society of New Zealand*, vol. 37, 2007, pp. 59–73.
2   J. W. Stack, *Kaiapohia: The Story of a Siege*, Whitcombe & Tombs, Christchurch, 1893.
3   E. Best, *The Astronomical Knowledge of the Maori*, Monograph 3, Dominion Museum, Wellington, 1922.
4   E. Best, *The Maori as He Was: A Brief Account of Maori Life as it Was in Pre-European Days*, New Zealand Board of Science and Art, Manual 4, Dominion Museum, Wellington, 1934.
5   See J. Williams, 'E Pākihi Hakinga a Kai', PhD thesis, University of Otago, 2004, for an account of mahika kai in Te Wai Pounamu. Aspects of environmental learning in early colonial New Zealand are discussed in P. Holland, J. Williams and V. Wood, 'Learning About the Environment in Early Colonial New Zealand', in T. Brooking and E. Pawson (eds), *Seeds of Empire: The Environmental Transformation of New Zealand*, I. B. Taurus, London, 2011, pp. 34–50.
6   D. Cresswell, *The Story of Cheviot*, Cheviot County Council, Hawarden, 1951, p. 43.
7   A. J. Davey, *Daybreak in Geraldine County, 1877–1952*, privately published, Geraldine, 1952, p. 21.
8   J. H. Beattie, *Traditions and Legends of the South Island Maori: Collected from Natives of Murihiku (Southland, New Zealand)*, text drawn from issues of the *Journal of the Polynesian Society* (1915 to 1922), collated by A. Anderson and republished by Cadsonbury Publications, Christchurch, 2004.
9   P. Bawden, *The Years Before Waitangi: A Story of Early Maori / European Contact in New Zealand*, published by the author, Auckland, 1987, p. 21.
10   J. Ward, *Information Relative to New Zealand, Compiled for the Use of Colonists*, second edition, John W. Parker, London, 1840, p. 67.
11   R. A. Cruise, *Journal of a Ten Months' Residence in New Zealand*, Longman, Hurst, Rees, Orme, Brown & Green, London, 1823, reprinted Pegasus Press, Christchurch, 1957, p. 25.
12   Cruise, *Journal of a Ten Months' Residence*, p. 27.
13   Ward, *Information Relative to New Zealand*, p. 92.
14   A. D. McIntosh (ed.), *Marlborough: A Provincial History*, Marlborough Provincial Historical Committee, Blenheim, 1940, p. 117.
15   O. A. Gillespie, *South Canterbury: A Record of Settlement*, South Canterbury Centennial History Committee, Timaru, 1958, p. 51.
16   Ibid.
17   J. C. Andersen, *Jubilee History of South Canterbury*, Whitcombe & Tombs, Auckland, 1916.
18   E. Shortland, *The Southern Districts of New Zealand; A Journal with Passing Notices of the Customs of the Aborigines*, Longman, Brown, Green & Longman, London, 1851, p. 205.
19   Bawden, *The Years Before Waitangi*, p. 38.
20   E. Pawson and P. Holland, 'Lowland Canterbury Landscapes in the Making', *New Zealand Geographer*, vol. 61, 2005, pp. 167–75.
21   H. C. Evison, *Te Wai Pounamu the Greenstone Island: A History of the Southern Maori During the European Colonisation of New Zealand*, Aoraki Press, Wellington, 1993, p. 177.
22   Ibid.
23   Gillespie, *South Canterbury*, p. 57.
24   A. H. Clark, *The Invasion of New Zealand by People, Plants, and Animals: The South Island*, Rutgers University Press, New Brunswick, 1949, p. 120.
25   Thomas Ferens' diaries, Waikouaiti, east Otago (transcript): Hocken Collections. MS-0440/016.
26   James Williams, personal communication, 2011.
27   Gillespie, *South Canterbury*.

28 L. G. D. Acland, *The Early Canterbury Runs*, Whitcombe & Tombs, Christchurch, 1951, p. 250.

29 Gillespie, *South Canterbury*, pp. 74–75.

30 Evison, *Te Wai Pounamu*, pp. 381–82.

31 C. Hursthouse, *New Zealand, the 'Britain of the South'; With a Chapter on the Native War, and our Future Native Policy*, second edition, E. Stanford, London, 1861.

32 W. D. McIntyre (ed.), *The Journal of Henry Sewell, 1853–1857*, two volumes, Whitcoulls, Christchurch, 1980.

33 McIntosh (ed.), *Marlborough*, p. 303.

34 R. Speight, A. Wall and R. M. Laing (eds), *Natural History of Canterbury*, Simpson & Williams, Christchurch, 1927.

35 Evison, *Te Wai Pounamu*, pp. 402–3. In colonial times, the Māori word whata was occasionally spelled 'futtah': A. Wall, 'A Sheep Station Glossary', in Acland, *The Early Canterbury Runs*, pp. 355–411.

36 Andersen, *Jubilee History*.

37 K. J. Fearon, *Te Wharau*, Netherton Grange Publications, Masterton, 1993.

38 John Barnicoat's diaries, Nelson and eastern South Island (transcript): Hocken Collections, MS-1451.

39 J. C. Andersen, field book containing notes about Māori nomenclature and legends: Alexander Turnbull Library, Wellington, MS-PAPERS-0148-112.

40 J. Cowan, loose-leaf folder containing miscellaneous items: Alexander Turnbull Library, Wellington, MS-PAPERS-0039.

41 M. Gordon, *The Garden of Tane*, A. H. & A. W. Reed, Wellington, not dated.

42 Cruise, *Journal of a Ten Months' Residence*, p. 73.

43 Anon., *Letters from Settlers and Labouring Emigrants in the New Zealand Company's Settlements of Wellington, Nelson, and New Plymouth, from February 1842 to January 1843*, Smith, Elder & Co., London, 1843, letter 49, July 1842.

44 Francis Pillans' diaries, Inch Clutha, south Otago (transcript): Hocken Collections, MS-0616.

45 Hursthouse, *New Zealand*, p. 25.

46 Lady Mary Barker, *Station Life in New Zealand*, Whitcombe & Tombs, Christchurch, 1950, p. 156.

47 Lady Mary Broome, *Colonial Memories*, Smith, Elder & Co., London, 1904, p. 23.

48 Andersen, *Jubilee History*, p. 116

49 Andersen, field book, MS-0148-112.

50 James Herries Beattie, handwritten notes, 1915–1922: Hocken Collections, MS-0181.

51 Shortland, *The Southern Districts*, p. 235.

52 Andersen, *Jubilee History*.

53 Ward, *Information Relative to New Zealand*, p. 85.

54 Gordon, *The Garden of Tane*, p. 67.

55 J. W. Stack, edited by A. H. Reed, *A White Boy Among the Maoris in the 'Forties': Pages from an Unpublished Autobiography of James West Stack*, A. H. Reed, Dunedin, 1934.

56 E. M. L. Studholme, *Reminiscences of 1860*, Whitcombe & Tombs, Christchurch, not dated.

57 Broome, *Colonial Memories*.

58 E. C. Richards (ed.), *Diary of E. R. Chudleigh 1862–1921: Chatham Islands*, Simpson & Williams, Christchurch, 1950.

59 Speight, Wall and Laing (eds), *Natural History of Canterbury*.

60 Best, *The Maori as He Was*.

61 Fearon, *Te Wharau*.

62 Williams, 'E Pākihi Hakinga a Kai'.

63 Clark, *The Invasion of New Zealand*, p. 293.

64 Cruise, *Journal of a Ten Months' Residence*, p. 195.

65 Anon., 1843, Letter 47, July 1842.

66 J. R. Godley (ed.), *Letters from Early New Zealand by Charlotte Godley 1850–1853*, Whitcombe & Tombs, Christchurch, 1951.

67 Evison, *Te Wai Pounamu*.

68 Ibid.

69 Ward, *Information Relative to New Zealand*, pp. 91–92.

70 Ibid., p. 3.

71 Shortland, *The Southern Districts*, p. 165.

72 Gillespie, *South Canterbury*, p. 59.

73 McIntyre (ed.), *The Journal of Henry Sewell*, p. 437.

2 Settlers Learning about Wind, Warmth and Rain

1 H. S. Selfe, *Canterbury New Zealand in 1862*, G. Street, 'New Zealand Examiner' Office, London, 1862, pp. 8–9.

2 P. B. Maling (ed.), *The Torlesse Papers: The Journals and Letters of Charles Obins Torlesse Concerning the Foundation of the Canterbury Settlement in New Zealand*

1848–51, Pegasus Press, Christchurch, 1958.

3 E. R. Ward, *The Journal of Edward Ward: 16 December 1850 to 22 June 1851*, Pegasus Press, Christchurch, 1951.

4 W. D. McIntyre (ed.), *The Journal of Henry Sewell, 1853–1857*, two volumes, Whitcoulls, Christchurch, 1980.

5 P. Holland and W. Mooney, 'Wind and Water: Environmental Learning in Early Colonial New Zealand', *New Zealand Geographer*, vol. 62, 2006, pp. 40–50.

6 Ward, *The Journal of Edward Ward*, p. 4.

7 McIntyre (ed.), *The Journal of Henry Sewell*.

8 The Point diaries, mid-Canterbury: Canterbury Museum, Christchurch.

9 John Barnicoat's diaries, Nelson and eastern South Island (transcript): Hocken Collections, MS-1451.

10 McIntyre (ed.), *The Journal of Henry Sewell*.

11 Frederick W. Teschemaker's diaries, south Canterbury and east Otago (transcript): Hocken Collections, MS-0446.

12 S. Siegel, *Nonparametric Statistics for the Behavioural Sciences*, McGraw-Hill, New York, 1956.

13 E. C. Richards (ed.), *Diary of E. R. Chudleigh 1862–1921: Chatham Islands*, Simpson & Williams, Christchurch, 1950.

14 John Gunn's diaries, Hook, south Canterbury: Hocken Collections, MS-1070.

15 P. Holland, P. Dixon and V. Wood, 'Learning About the Weather in Early Colonial New Zealand', *Weather and Climate*, vol. 29, 2009, pp. 3–21.

16 'How to Foretell the Weather', *Southern Cross*, 14 December 1860. The published text was drawn from *Manual of the Barometer* by Robert FitzRoy, published by the Board of Trade, London.

17 K. Anderson, *Predicting the Weather: Victorians and the Science of Meteorology*, University of Chicago Press, Chicago, 2005, p. 149.

18 Ibid., pp. 105–30.

19 Ibid., pp. 46–55.

20 Ida Valley Station letter books, letter dated 21 July 1879: Hocken Collections, MS-0658.

21 William Shirres' letter book, Aviemore Station, Waitaki Valley, letter dated 20 June 1884: Hocken Collections, 291*MS-0635.

22 S. Grant and J. S. Foster, *New Zealand: A Report on its Agricultural Conditions and Prospects*, G. Street, London, 1880, p. 30.

23 William Shirres' letter book, Aviemore Station, Waitaki Valley, letter dated 1 October 1886: Hocken Collections, 291*MS-0635.

3 *Exceptional Challenges*

1 H. S. Selfe, *Canterbury New Zealand in 1862*, G. Street, 'New Zealand Examiner' Office, London, 1862, p. 9.

2 J. Bathgate, *New Zealand: Its Resources and Prospects*, W. & R. Chambers, London, 1880.

3 S. Grant and J. S. Foster, *New Zealand: A Report on its Agricultural Conditions and Prospects*, G. Street, London, 1880.

4 J. Hector, *Handbook of New Zealand*, Lyon & Blair, Wellington, 1880. The previous year, the same publishers had produced a handbook with an identical title for the Sydney International Exhibition.

5 Archibald Morton's letter book, Haldon Station, letters to J. H. Preston: Hocken Collections, MS-1272/064.

6 Archibald Morton's letter book, Haldon Station, letter to J. H. Preston dated 27 September 1898: Hocken Collections, MS-1272/064.

7 File of correspondence from J. Rattray & Son, Dunedin to J. H. Preston, Longlands Station: Hocken Collections, MS-1272/105.

8 Outward correspondence from the Gore to the Invercargill branches of the New Zealand Loan & Mercantile Agency: Hocken Collections, AG-875/006.

9 William Shirres' letter book, Aviemore Station, Waitaki Valley, letter dated 14 March 1886: Hocken Collections, 291A*MS-0635.

10 Archibald Morton's letter book, Haldon Station, letters to J. H. Preston: Hocken Collections, MS-1272/064.

11 William Shirres' letter book, Aviemore Station, Waitaki Valley, letter dated 3 October 1884: Hocken Collections, 291A*MS-0635.

12 Ida Valley Station letter books and diaries,

Central Otago: Hocken Collections,
MS-0658.

13 Application by J. H. Preston to
the Chairman of the Land Board,
Christchurch, for relief under The
Pastoral Tenants Act 1895: Hocken
Collections, MS-1271/016.

14 Archibald Morton's letter book, Haldon
Station, letters to J. H. Preston: Hocken
Collections, MS-1272/064.

15 The information presented in Figure 3.1
was collated from the following runs of
farm and station diaries and letter books:
Thomas and William Adams' diaries
(selections transcribed by J. H. Beattie),
Waihola, south Otago: Hocken
Collections, MS-582
Bryce Brothers diaries (most entries by
David Bryce), Lovells Flat, south Otago:
Hocken Collections, MS-0615
Edward Chudleigh, mid-Canterbury: E. C.
Richards (ed.), Diary of E. R. Chudleigh
1862–1921: Chatham Islands, Simpson
& Williams, Christchurch, 1950
Joseph Davidson's diaries, Switzers /
Waikaia, northern Southland: Hocken
Collections, AG-523
John Hall's diaries, Edendale, Southland:
Hocken Collections, MS-0292
Ida Valley Station letter books and diaries,
Central Otago: Hocken Collections,
MS-0658
John Maddison's diaries, Southbridge-
Leeston, mid-Canterbury: Canterbury
Museum, Christchurch, ARC 1900.325
McMaster family diaries, Tokarahi,
Waitaki Valley, north Otago: Hocken
Collections, MS-1011
James Menzies' diaries, Mataura,
Southland: Hocken Collections
(transcript), MS-3178/053
Mount Nicholas Station diaries, Lake
Wakatipu basin (partial transcript):
Hocken Collections, MS-1270-2-3/002
Joseph Munnings' diaries, Christchurch
and environs (transcript): Canterbury
Museum, Christchurch, ARC 1990.50,
item 4
James Murison's diaries, Puketoi Station,
Maniototo, eastern Otago (transcript):
Hocken Collections, ARC-0359
Francis Pillans' diaries, Inch Clutha, south
Otago (transcript): Hocken Collections,
MS-0616

James Preston's diaries, letter books and
personal papers, Southland, eastern
Otago, Maniototo, Waitaki Valley
and Mackenzie Country: Hocken
Collections, MS-1271 and MS-1272
The Point diaries, mid-Canterbury:
Canterbury Museum, Christchurch
John Wither's diaries, Lake Wakatipu
basin, Central Otago (transcript):
Hocken Collections, 89-149.

4   Away with the Old

1   J. C. Firth, 'On Forest Culture', New
Zealand Country Journal, vol. 3, 1879,
pp. 142–49.

2   S. Grant and J. S. Foster, New Zealand:
A Report on its Agricultural Conditions and
Prospects, G. Street, London, 1880, p. 59
and pp. 53–54, respectively.

3   Joseph Munnings' diaries, Christchurch
and environs (transcript): Canterbury
Museum, Christchurch, ARC 1990.50,
item 4.

4   E. Pawson and P. Holland, 'People and
the Land', in M. Winterbourn, G. Knox,
C. Burrows and I. Marsden (eds), The
Natural History of Canterbury, third
edition, Canterbury University Press,
Christchurch, 2008, Figure 1 on p. 38.

5   The Point diaries, mid-Canterbury:
Canterbury Museum, Christchurch.

6   Macdonald family diaries, Orari Station,
south Canterbury: in private hands.

7   Grampian Hills Station diaries,
Mackenzie Country (transcript): South
Canterbury Museum, Timaru.

8   Kate Sheath's diaries, Albury, south
Canterbury (typed transcript): Waimate
Historical Society and Museum.
The handwritten original is in the South
Canterbury Museum, Timaru.

9   Native birds were of interest to collectors
in England, and a short piece published
in the 27 March 1869 issue of the Timaru
Herald noted the recent export of pūkeko,
paradise and blue ducks, and two native
falcons.

10 Frederick and Thomas Teschemaker's
diaries, Otaio Station, south Canterbury:
South Canterbury Museum, Timaru.
The settlers' word 'snig' meant 'to drag
along the ground by horse and bullocks,

especially to drag logs or other timber':
A. Wall, 'A Sheep Station Glossary', in
L. G. D. Acland, *The Early Canterbury
Runs*, Whitcombe & Tombs, Christchurch,
1951, pp. 355–411.

11 Raincliff Station diaries, south Canterbury
(transcript): South Canterbury Museum,
Timaru.

12 J. H. Simmonds, 'Wind-Breaks for
Fruit Farms', *New Zealand Journal of
Agriculture*, vol. 15, 1917, pp. 253–62.

13 The Waimate Historical Society and
Museum holds an incomplete run of ledgers relating to Te Waimate Station. They
were maintained by the now defunct
company Manchester and Goldsmith, and
while they do not provide a complete account of annual income and expenditure
for several years between 1876 and 1888,
they appear comprehensive for outlays
and income for shearing, ploughing,
harvesting, fencing (including materials,
installation and maintenance), transport
of products to markets, and the range of
contract labour employed on this large
pastoral property.

14 N. C. Clifton, *New Zealand Timbers: The
Complete Guide to Exotic and Indigenous
Woods*, GP Publications, Wellington, 1991,
pp. 78–79.

15 A. E. Green, 'Native Trees and Shrubs as
Hedge Plants', *New Zealand Journal of
Agriculture*, vol. 4, 1912, pp. 444–48.

16 Ibid.

17 T. H. Potts, 'The Kea – Progress of
Development', *Nature*, vol. 4, 1871, p. 489;
and T. H. Potts, 'Change of Habits in
Animals and Plants', *Nature*, vol. 5, 1872,
p. 262.

18 S. S. Crawford, *Sheep and Sheepmen
of Canterbury, 1850–1914*, Simpson &
Williams, Christchurch, 1949.

19 John Finlay's diaries, Makikihi, south
Canterbury (transcript): Waimate
Historical Society and Museum.

20 Cody family diaries, Lime Hills, northern
Southland: Hocken Collections, AG-736.

21 Kauru Hill diaries, north Otago: North
Otago Museum, Oamaru, Box 59/33d.

22 James Menzies' diaries, Mataura,
Southland (transcript): Hocken
Collections, MS-3178/053.

23 Joseph Munnings' diaries, Christchurch
and environs (typed transcript):

Canterbury Museum, Christchurch, ARC
1990.50.

24 H. S. Selfe, *Canterbury New Zealand in
1862*, G. Street, 'New Zealand Examiner'
Office, London, 1862.

25 W. Cronon, 'The Trouble With
Wilderness: or, Getting Back to the Wrong
Nature', *Environmental History*, vol. 1,
1995, pp. 7–55.

26 L. Cockayne, 'An Economic Investigation
of the Montane Tussock-Grassland of
New Zealand. No. I – Introduction', *New
Zealand Journal of Agriculture*, vol. 18,
1919, pp. 1–9.

27 William Shirres' letter book for Aviemore
Station, Waitaki Valley, Hocken
Collections, 291A*MS-0635.

28 John Finlay's diaries, Makikihi, south
Canterbury (typed transcript): Waimate
Historical Society and Museum.

29 W. P. Reeves, *New Zealand, and Other
Poems*, Grant Richards, London, 1898, p. 8.

30 W. P. Reeves, *New Zealand*, Adam &
Charles Black, London, 1908, p. 103.

31 J. F. Armstrong, 'On the Naturalized
Plants of the Province of Canterbury',
Transactions of the New Zealand Institute,
vol. 4, 1871, pp. 284–90.

32 J. B. Armstrong, 'The Native Grasses',
New Zealand Country Journal, vol. 3, 1879,
pp. 201–4.

33 Manager's letter book, New Zealand and
Australian Land Company, Levels Station,
south Canterbury: South Canterbury
Museum, Timaru.

34 R. Wilkin, 'A Gossip About Grasses', *New
Zealand Country Journal*, vol. 3, 1879,
pp. 294–98.

35 M. Dixon, 'Permanent Pasture Grasses',
New Zealand Country Journal, vol. 10,
1886, pp. 35–38.

36 J. B. Armstrong, 'The Native Grasses',
p. 203.

37 J. Buchanan, *Manual of the Indigenous
Grasses of New Zealand*, Colonial
Museum, Wellington, 1880.

38 E. Edgar and H. E. Connor, *Flora of New
Zealand. Volume V, Graminae*, Manaaki
Whenua Press, Lincoln, 2000.

39 J. G. Wilson, 'Grasses', *New Zealand
Farmer, Bee and Poultry Journal*, vol. 21,
1901, pp. 1–3.

40 J. B. Armstrong, 'The Native Grasses',
p. 448.

41   Crawford, *Sheep and Sheepmen*.
42   W. H. Taylor, 'Native Plants: The Work of Collection and Establishment', *New Zealand Journal of Agriculture*, vol. 7, 1913, pp. 61–64.
43   H. C. Jacobson and J. W. Stack, *Tales of Banks Peninsula*, third edition, Akaroa Mail, Akaroa, 1917, p. 324.
44   J. W. Stack, *Through Canterbury and Otago with Bishop Harper in 1858–60*, Akaroa Mail, Akaroa, not dated, p. 19.
45   J. M. Hopkins, 'Native Bush and Excellent Stock Shelter: Its Protection and Regeneration Well Worth While', *New Zealand Journal of Agriculture*, vol. 75, 1975, pp. 257–59.

5   In with the New

1   J. W. Stack, *Kaiapohia: The Story of a Siege*, Whitcombe & Tombs, Christchurch, 1893, p. 10.
2   A. H. Cockayne, 'The Grasslands of New Zealand: Component Species', *New Zealand Journal of Agriculture*, vol. 17, 1918, pp. 210–20. Twelve years later that remarkable observer, Herbert Guthrie-Smith, who owned Tutira Station in Hawke's Bay, documented successional change in sown pasture on his property: H. Guthrie-Smith, 'Grassland Evolution at Tutira', *New Zealand Journal of Agriculture*, vol. 40, 1930, pp. 84–92.
3   E. B. Levy, 'Observations Relative to Pasture Seed-Mixtures', *Proceedings of the Fifth Conference of the New Zealand Grasslands Association*, 1936, pp. 33–40.
4   P. D. Sears, 'Exploitation of High Production Pastures in New Zealand', *Proceedings of the Ninth Conference of the New Zealand Ecological Society*, 1962.
5   A. H. Cockayne, 'The Surface-Sown Grass Lands of New Zealand', *New Zealand Journal of Agriculture*, vol. 7, 1913, pp. 465–75.
6   B. L. Evans, 'Grassland Research in New Zealand', *New Zealand Official Year-Book*, Government Printer, Wellington, 1960, pp. 1234–64.
7   Anon., 'Mixtures of Grass Seed', *New Zealand Stock and Station Journal*, June 1902, p. 269.
8   T. Kjægaard, 'A Plant that Changed the World: The Rise and Fall of Clover, 1000–2000', *Landscape Research*, vol. 28, 2003, pp. 41–49.
9   E. B. Levy, 'The Grasslands of New Zealand: Principles of Pasture Establishment', *New Zealand Journal of Agriculture*, vol. 23, 1921(a), pp. 259–65.
10   Ibid., pp. 321–30.
11   P. Patullo, 'Farming Experiments', *New Zealand Country Journal*, vol. 22, 1898, pp. 374–79.
12   O. G. Parker, 'Laying Down Land to Grass and its Subsequent Treatment', *New Zealand Country Journal*, vol. 5, 1881, pp. 401–4. The following decade, a pamphlet was commercially published on the topic of establishing permanent pastures. The author aimed 'to place before my readers information which I think will be of service to them': G. C. Tothill, *Laying Down Land in Permanent Grasses*, Southland Times Company, Invercargill, 1893.
13   A. Simson, 'Pasturage', *New Zealand Country Journal*, vol. 10, 1886, pp. 291–95.
14   J. G. Wilson, 'Grasses', *New Zealand Farmer, Bee and Poultry Journal*, vol. 20, 1900, pp. 457–58.
15   Letter book, Ida Valley Station, Central Otago: Hocken Collections, MS-0658/027.
16   James Preston's invoices and receipts relating to Centrewood Station, east Otago: Hocken Collections, MS-1272/090.
17   R. Wilkin, 'Grasses and Forage Plants Best Adapted to New Zealand', *New Zealand Country Journal*, vol. 1, 1877, pp. 3–12.
18   'Ovis', 'Permanent Pastures and Alternative Farming', *New Zealand Country Journal*, vol. 13, 1889, pp. 451–56.
19   A. H. Cockayne, 'Surface-Sown Grass Mixtures: Some Recent Changes', *New Zealand Journal of Agriculture*, vol. 14, 1917, pp. 169–77.
20   A. H. Cockayne, 'The Grasslands of New Zealand: Component Species'.
21   In nineteenth-century newspaper advertisements, it was referred to as 'perennial ryegrass', and not simply to differentiate it from 'Italian ryegrass', but farmers found that what had been sold to them as 'perennial ryegrass' was often a short-lived form.
22   A. H. Cockayne, 'The Grasslands of New Zealand: Component Species'.

23   Simson, 'Pasturage'.
24   'Ovis', 'Pastures', *New Zealand Country Journal*, vol. 21, 1897, pp. 393–97.
25   A. H. Cockayne, 'Surface-Sown Grass Mixtures'.
26   R. Wilkin, 'A Gossip about Grasses', *New Zealand Country Journal*, vol. 3, 1879, pp. 294–98.
27   'Ovis', 'Pastures'.
28   P. McConnell, 'Cultivation', *Journal of the Department of Agriculture*, 15 December 1911, pp. 448–53.
29   Anon., 'Permanent Pastures', *New Zealand Farmer, Stock and Station Journal*, vol. 27, 1906, pp. 713–15.
30   A. H. Cockayne, 'The Grasslands of New Zealand: Component Species'.
31   Frederick and William Teschemaker's diaries, Otaio Station, south Canterbury (transcript): South Canterbury Museum, Timaru.
32   John Finlay's diaries, Makikihi, south Canterbury (transcript): Waimate Historical Society and Museum.
33   Anon., 'The Farm; Putting Down to Grass', *New Zealand Farmer, Stock and Station Journal*, vol. 34, 1913, pp. 37–38.
34   A. H. Cockayne, 'The Grasslands of New Zealand'.
35   J. M. Smith, 'Some Aspects of the Extreme Simplification of Pasture Seed-Mixtures', *Proceedings of the Fifth Conference New Zealand Grasslands Association*, 1936, pp. 41–45.
36   R. G. S., 'Seed Mixtures: Controversies Past, Present and Future', in I. H. Hunter (ed.), *Bailliere's Encyclopaedia of Scientific Agriculture*, Bailliere, London, 1931, pp. 1104–18.
37   A. H. Cockayne, 'The Grasslands of New Zealand: Component Species'.
38   A. H. Cockayne, 'Surface-Sown Grass Mixtures'.
39   E. Gray, 'The Early History of Ryegrass Seed Production in Scotland', *Seed Trade Review*, April 1952, pp. 103–10.
40   F. W. Hutton, 'Cross-Fertilisation of Plants and its Relation to Agriculture', *New Zealand Country Journal*, vol. 7, 1883, pp. 331–34.
41   J. C. Crawford, 'On Hedges and Hedge Plants', *Transactions of the New Zealand Institute*, vol. 9, 1876, pp. 203–6.
42   A. E. Woodhouse, *George Rhodes of the Levels and his Brothers*, Whitcombe & Tombs, Auckland, 1937.
43   J. R. Godley (ed.), *Letters from Early New Zealand by Charlotte Godley 1850–1853*, Whitcombe & Tombs, Christchurch, 1951.
44   'Agricola', 'Hedge Plants', *New Zealand Country Journal*, vol. 1, 1877, pp. 94–95.
45   J. S. Yeates, 'Farm Trees and Hedges', *Massey Agricultural College Bulletin*, no. 12, 1948.
46   Peter Nelson, County Clerk to J. M. McKenzie, Member of the House of Representatives, Clutha County outward letter book, 1879–1881, letter dated 1 May 1889: Hocken Collections, 323*AG-253-001/001.
47   See Note 13, Chapter 4: the Waimate Historical Society and Museum holds a large, but incomplete, run of ledgers relating to Te Waimate Station.
48   Frederick and William Teschemaker's diaries, Otaio Station, south Canterbury: South Canterbury Museum, Timaru.
49   P. Holland and P. Dixon, 'New Plants for the "Howling Wilderness"', *Bulletin of the Dunedin Rhododendron Group*, no. 34, 2006, pp. 18–29.
50   C. Siebert, 'Food Ark', *National Geographic*, vol. 221, 2011, pp. 108–31.
51   W. D. McIntyre (ed.), *The Journal of Henry Sewell, 1853–1857*, two volumes, Whitcoulls, Christchurch, 1980.
52   The Point diaries, mid-Canterbury: Canterbury Museum, Christchurch.
53   C. A. Sharp (ed.), *The Dillon Letters: The Letters of the Hon. Constantine Dillon, 1842–1853*, A. H. & A. W. Reed, Wellington, 1954.
54   P. Holland and A. Wearing, 'By Choice: Plants Grown by Settlers in Lowland Canterbury', in M. Roche, M. McKenna and P. Hesp (eds), *Proceedings of the Twentieth New Zealand Geographical Society Conference*, New Zealand Geographical Society, Hamilton, 1999, pp. 50–54.
55   Sharp (ed.), *The Dillon Letters*.
56   S. Challenger, 'Pioneer Nurserymen of Canterbury, New Zealand (1860–65)', *Garden History*, vol. 7, 1979, pp. 25–64.
57   James Miller's diaries. Outram–Maungatua, east Otago: Hocken Collections, 91-010.

58 B. Wilson, 'Stock to Fit the Environment', *New Zealand Farmer*, August 1982, pp. 184–99 and 201–7.

59 S. Grant and J. S. Foster, *New Zealand: A Report of its Agricultural Conditions and Prospects*, G. Street, London, 1880.

60 J. Hector, *Handbook of New Zealand*, Lyon & Blair, Wellington, 1880.

61 E. C. Richards (ed.), *Diary of E. R. Chudleigh 1862–1921: Chatham Islands*, Simpson & Williams, Christchurch, 1950.

62 The Point diaries, mid-Canterbury: Canterbury Museum, Christchurch.

63 Manchester and Goldsmith ledgers for Te Waimate Station, south Canterbury (see note 47).

64 William Shirres' letter book, Aviemore Station, Waitaki Valley, letter dated 10 September 1865: Hocken Collections, 291A*MS-0635.

65 Wilson, 'Stock to Fit the Environment'.

66 Ibid.

67 E. B. Levy, 'The Grasslands of New Zealand: Preliminary Ecological Classification of Species', *New Zealand Journal of Agriculture*, vol. 30, 1925, pp. 357–74.

6 Emerging Environmental Problems

1 S. Grant and J. S. Foster, *New Zealand: A Report on its Agricultural Conditions and Prospects*, G. Street, London, 1880, p. 70.

2 Ibid., p. 45.

3 Ibid., p. 40.

4 Ibid., p. 45.

5 Ibid., p. 46.

6 H. Guthrie-Smith, *Tutira: The Story of a New Zealand Sheep Station*, Blackwood, Edinburgh, 1953.

7 J. Bathgate, *New Zealand: Its Resources and Prospects*, W. & R. Chambers, London, 1880, pp. 15 and 25, respectively.

8 Joseph Munnings' diaries, Christchurch and environs (transcript): Canterbury Museum, Christchurch, ARC 1990.50, item 4.

9 Elsdon Best reported that Māori loosened the ground with a kō, pulverised clods, picked out and discarded roots, and used fine gravel to improve the texture of clay soils: *The Maori As He Was: A Brief Account of Maori Life as it Was in Pre-European Days*, New Zealand Board of Science and Art, Manual 4, Dominion Museum, Wellington, 1934. J. H. Beattie's informants told him that 'The old Maoris never manured the land – they thought it useless and scorned the Pakeha's way as stupid': 'Record of Interviews with Maori in Canterbury – Section XXVIII', Hocken Collections, MS-0181/005.

10 E. C. Richards (ed.), *Diary of E. R. Chudleigh, 1862–1921: Chatham Islands*, Simpson & Williams, Christchurch, 1950.

11 Grant and Foster, *New Zealand*, p. 6.

12 *New Zealand Official Yearbook*, Government Printer, Wellington, 1894, p. 276.

13 Small tubers tend to indicate nutrient deficiency; he did not refer to small shoots, which would suggest a shortage of water.

14 R. Wilkin, 'Does the Continuous Grazing of Land with Sheep Cause it to Deteriorate in Value?', *New Zealand Country Journal*, vol. 5, 1881, pp. 3–4. An answer to that question was provided by the Australian analytical chemist H. W. Potts, who reported American research which showed 'that a ton of butter will remove fertility from the soil valued at 2s. 6d. above what the cow returns, whilst the fertility used to produce a ton of wheat would cost 29s. to replace': 'The Exhaustion of our Grazing Lands', *New Zealand Country Journal*, vol. 21, 1897, pp. 507–9.

15 Outward correspondence from the manager of Gore branch of the New Zealand Loan & Mercantile Agency to the manager of the company's office in Invercargill, letter dated 6 March 1905: Hocken Collections, AG-875/006.

16 H. Guthrie-Smith, 'Grassland Evolution at Tutira', *New Zealand Journal of Agriculture*, vol. 40, 1930, pp. 84–92.

17 E. B. Levy, 'The Grasslands of New Zealand: Preliminary Ecological Classification of Species', *New Zealand Journal of Agriculture*, vol. 30, 1925, pp. 357–74.

18 Macdonald family diaries, Orari Station, south Canterbury: in private hands.

19 K. A. Wodzicki, *Introduced Mammals of New Zealand: An Ecological and Economic Survey*, Bulletin No. 98, Department of Scientific and Industrial Research,

Wellington, 1950. Figure 18 on p. 108 shows 1862 as the year when rabbits were introduced in southern New Zealand, but entries in farm and station diaries suggest at least a decade earlier.

20 P. Holland, K. O'Connor and A. Wearing, 'Remaking the Grasslands of the Open Country', in E. Pawson and T. Brooking (eds), *Environmental Histories of New Zealand*, Oxford University Press, Melbourne, 2002, pp. 69–83.

21 Richards (ed.), *Diary of E. R. Chudleigh*.

22 W. H. McLean, *Rabbits Galore – on the Other Side of the Fence*, Reed, Wellington, 1966, p. 96.

23 T. G. Sewell, 'Improvement of Tussock Grassland', *New Zealand Journal of Agriculture*, vol. 81, 1950, pp. 503–17.

24 'Report of the Rabbit Nuisance Committee', *Appendix to the Journals of the House of Representatives of New Zealand*, I-5, 1886.

25 Ibid.

26 'Inoculation of Rabbits', *Appendix to the Journals of the House of Representatives of New Zealand*, H-11, 1887.

27 'The Rabbit Nuisance: Annual Report by the Superintending Inspector', *Appendix to the Journals of the House of Representatives of New Zealand*, H-18, 1888.

28 'The Supply of Ferrets', *Appendix to the Journals of the House of Representatives of New Zealand*, H-31, 1888.

29 'Lands Thrown-up in Rabbit-Infested Districts in Otago and Southland', *Appendix to the Journals of the House of Representatives of New Zealand*, C-17, 1887.

30 S. S. Crawford, *Sheep and Sheepmen of Canterbury, 1850–1914*, Simpson & Williams, Christchurch, 1949.

31 William Shirres' letter book, Aviemore Station, Waitaki Valley, letter dated 20 June 1884: Hocken Collections, 291A*MS-0635.

32 Letter dated 22 November 1904 from William Grant to J. H. Preston: Hocken Collections, MS-1272/066.

33 Correspondence relating to J. H. Preston's Haldon Station, Mackenzie Country: Hocken Collections, MS-1272/096.

34 Ibid.

35 Archibald Morton's letter book for Haldon Station, Mackenzie Country, letter dated 12 July 1898 to J. H. Preston: Hocken Collections, MS-1272/064.

36 Letter by C. C. Boyes to the editor of the *Otago Witness*, 23 October 1880.

37 The manager of Aviemore Station informed the financial backer in London that when making a batch of phosphorised Sparrowbill oats he dissolved phosphorus sticks in a hot salt or sugar solution, but preferred the former. William Shirres' letter books, Aviemore Station, letter dated 19 October 1885: Hocken Collections, 291*MS-0635.

38 Archibald Morton's letter book for Haldon Station, Mackenzie Country: Hocken Collections, MS-1272/064.

39 The shearing accounts book for James Preston's Haldon Station, Mackenzie Country included an entry dated 25 April 1899 for the purchase of '9 bags of poison sugar @ 70lb per bag' and '3 tins of phosphorus': Hocken Collections, MS-0989/247. There were few explicit references to 'poison sugar' in Preston's diaries for Haldon and Black Forest stations, which were effectively operated as one property, but he was still using it fourteen years later ('I took pollard and poison sugar to Black Forest', entry dated 6 September 1913): Hocken Collections, MS-0989/279.

40 Cody family diaries, Lime Hills, northern Southland: Hocken Collections, AG-736.

41 McLean, *Rabbits Galore*.

42 Ibid., p. 87.

43 Ibid., p. 152

44 Ibid., p. 107.

45 Ramsay family diaries, Hyde, on the fringes of Central Otago: Hocken Collections, AG-680.

46 A. H. Cockayne, 'The Grasslands of New Zealand', *New Zealand Journal of Agriculture*, vol. 17, 1918, pp. 140–42.

47 Nugent Wade's letter books, Taipo Hill, Maheno, north Otago: South Canterbury Museum, Timaru.

48 Kauru Hill diaries, north Otago: North Otago Museum, Box 59/33d.

49 A. H. Cockayne, 'Impure Seed: The Source of our Weed Problem', *New Zealand Journal of Agriculture*, vol. 4, 1912, pp. 437–43.

50 A. H. Cockayne, 'Seed Impurities', *New Zealand Journal of Agriculture*, vol. 13, 1916, pp. 208–12.

51 Anon., 'The Purity of Seed', *New Zealand Farmer*, vol. 25, 1904, pp. 964–65.

52 Grant and Foster, *New Zealand*, p. 20.

53 'Petition of Inhabitants of Canterbury for Repeal of the Thistle Ordinance', *Appendix to the Journals of the House of Representatives of New Zealand*, F-9, 1870.

54 Nugent Wade's letter books, south Canterbury and north Otago: South Canterbury Museum, Timaru.

55 Kauru Hill diaries, north Otago: North Otago Museum, Oamaru.

56 Macdonald family diaries, Orari Station, south Canterbury: in private hands.

57 Ramsay family diaries, Hyde, on the fringes of Central Otago: Hocken Collections, AG-680.

58 Cody family diaries, Lime Hills, northern Southland, Hocken Collections, AG-736.

59 James Miller's diaries, Outram–Maungatua, east Otago: Hocken Collections, 91-010.

60 Manager's letter books, Australia New Zealand Land Company, south Canterbury: South Canterbury Museum, Timaru.

7 Opportunities to See, Hear and Compare

1 P. Holland and V. Wood, 'Decorative Plants and Horticultural Exhibitions in Early New Zealand', *Bulletin of the Dunedin Rhododendron Group*, no. 36, 2008, pp. 15–28.

2 During the second half of the nineteenth century, newspaper readers were kept informed by monthly or seasonal gardening calendars that had been written by local people with local conditions in mind.

3 L. de Bray, *Manual of Old-Fashioned Flowers*, Oxford Illustrated Press, Sparkford, 1984.

4 Holland and Wood, 'Decorative Plants and Horticultural Exhibitions'. One of Queen Victoria's children, who had recently toured New Zealand, was wounded by an Irish-born resident of Sydney. The Prince recovered, but the gunman was condemned to death. The affair sparked considerable public indignation in Christchurch and the Canterbury Horticultural Society saw this as an opportunity to demonstrate its allegiance to the Crown.

5 M. A. Morris, 'History of Christchurch Home Gardening from Colonisation to the Queen's Visit', PhD thesis, University of Canterbury, 2006.

6 Joseph Munnings' diaries, Christchurch and environs: Canterbury Museum, Christchurch, ARC 1990.50.

7 G. T. Vincent, 'Sports and Other Signs of Civilisation in Colonial Canterbury, 1850–1890', PhD thesis, University of Canterbury, 2002.

8 Cody family diaries, Lime Hills, northern Southland: Hocken Collections, AG-736.

9 J. Hector, *Handbook of New Zealand*, Lyon & Blair, Wellington, 1880, p. 14.

10 Anon., 'The First Pastoral and Agricultural Show', *New Zealand Country Journal*, vol. 1, 1877, pp. 21–26.

11 Anon., 'Judging at Shows', *New Zealand Country Journal*, vol. 13, 1889, pp. 456–58.

12 The information plotted in Figure 7.1 was collated from the following runs of farm and station diaries and letter books, listed in chronological order:
John Wither, Sunnyside, Lake Wakatipu basin, Central Otago (1871–1890): Hocken Collections, 89-149
Ida Valley Station, Central Otago (1871–1895): Hocken Collections, MS-0658
Bryce Brothers diaries (most entries by David Bryce) Lovells Flat, south Otago (1874–1900): Hocken Collections, MS-0615
Joseph Davidson, Switzers / Waikaia, northern Southland (1878–1900): Hocken Collections, AG-523
James Preston, Longlands Station, Maniototo, east Otago (1883–1890): Hocken Collections, MS-1271
McMaster family, Tokarahi, north Otago (1885–1893): Hocken Collections, MS-1011
Ramsay family, Hyde, Central Otago (1894–1925): Hocken Collections, AG-680
Cody family, Lime Hills, northern Southland (1901–1916): Hocken Collections, AG-736
James Miller, Outram-Maungatua, east Otago (1902–1910): Hocken Collections, 91-010.

13 Dickson Jardine's diaries, Otekaike, north Otago: Hocken Collections, AG-659.

14 P. Holland, P. Star and V. Wood, 'Pioneer Grassland Farming: Pragmatism, Innovation and Experimentation', in T. Brooking and E. Pawson (eds), *Seeds of Empire: The Environmental Transformation of New Zealand*, I. B. Tauris, London, 2011, pp. 51–72.

8   Rural People Continuing to Learn about their Environments

1   E. C. Semple, *Influences of the Geographical Environment, on the Basis of Ratzel's System of Anthropo-Geography*, H. Holt, New York, 1911.

2   E. C. Relph, *Place and Placelessness*, Pion, London, 1976; and Yi-fu Tuan, *Space and Place: The Perspective of Experience*, University of Minnesota Press, Minneapolis, 1977.

3   R. J. Johnson, 'There's a Place for Us', *New Zealand Geographer*, vol. 44, 1988, pp. 8–13.

4   M. Dominy, *Calling the Station Home: Place and Identity in New Zealand's High Country*, Rowman & Littlefield, Lanham, 2001.

5   J. Liebig, *Letters on Modern Agriculture*, Walton & Maberly, London, 1859.

6   S. Nathan and M. Varnham (eds), *The Amazing World of James Hector*, Awa Press, Wellington, 2008, p. 123.

7   A. G. Tansley, *Introduction to Plant Ecology*, G. Allen, London, 1946.

8   H. H. Barrows, 'Geography as Human Ecology', *Annals of the Association of American Geographers*, vol. 13, 1923, pp. 1–14.

9   J. Lovelock, *Gaia: A New Look at Life on Earth*, Oxford University Press, Oxford, 1987; and J. Lovelock, *Gaia: The Practical Science of Planetary Medicine*, Allen & Unwin, Sydney, 1991.

10 D. Dueck, *Strabo of Amasia*, Routledge, London, 2000.

11 A. von Humboldt (translated by E. C. Otte), *Cosmos: A Sketch of a Physical Description of the Universe*, five volumes, Harper, New York, 1870–1871.

12 J. C. Beaglehole (ed.), *The Endeavour Journal of Joseph Banks*, second edition, Public Library of New South Wales and Angus & Robertson, Sydney, 1963.

13 R. L. Peden, *Making Sheep Country: Mt Peel Station and the Transformation of the Tussock Lands*, Auckland University Press, Auckland, 2011.

14 Ida Valley Station letter books and diaries, Central Otago: Hocken Collections, MS-0658.

15 W. D. McIntyre (ed.), *The Journal of Henry Sewell, 1853–1857*, two volumes, Whitcoulls, Christchurch, 1980.

16 P. G. Holland, 'Cultural Landscapes as Biogeographical Experiments: A New Zealand Perspective', *Journal of Biogeography*, vol. 27, 2001, pp. 39–43.

17 K. B. Cumberland, *Soil Erosion in New Zealand: A Geographic Reconnaissance*, Whitcombe & Tombs in collaboration with the Soil Conservation and Rivers Control Council, Wellington, 1947.

18 Ministry for the Environment, *Environment New Zealand 2007*, Ministry for the Environment, Wellington, 2007.

19 T. H. Potts, *Out in the Open: A Budget of Scraps of Natural History Gathered in New Zealand*, Lyttelton Times Company, Christchurch, 1882. (This book is a compilation of articles previously published in the *New Zealand Country Journal*.)

20 M. Nicholson, *The Environmental Revolution: A Guide for the New Masters of the World*, Pelican, Harmondsworth, 1972, originally published by Hodder & Stoughton, London in 1970.

21 The following statements from farm and station diaries, letter books and correspondence relate to the model depicted in Figure 8.1.
    Local scale: 'Not a drop of water in the creeks yet. It seems to have been raining heavily all round lately, but a curse seems to rest on this valley.' (Manager of Ida Valley Station, Central Otago to A. D. Bell, letter dated 15 March 1888: Hocken Collections, MS-0658/027.)
    Regional scale: 'The wet weather is causing great delay and trouble everywhere down country. I believe at Rickland's Station only 8000 ewes were shorn in nearly three weeks. In this district although there has been so much rain, there has not been too much, I mean for the very dry country about here.' (Manager of Ida Valley

Station, Central Otago to Sir Frederick
Bell, letter dated 2 January 1877: Hocken
Collections, MS-0658/025.)
Larger scale: 'The country below Dunedin
is not so good, the flat parts are wet and full
of creeks. The climate is undoubtedly very
inferior, there is such a prevalence of SW
winds and there are no high mountains
like in Canterbury to break them and
draw the rain away, the breezes come off
the Pacific un-interruptedly and favour
country with more moisture than is agreeable. In fact it is very wet and cold both
overhead and underfoot.' (Diary of F. W.
Teschemaker, south Canterbury, eastern
Otago and Southland (transcript): Hocken
Collections, MS-0446.)
International scale: 'They are having at
home [i.e. England] an unusually mild
winter, which is often a presage of a
similar winter here [in southern New
Zealand]. Let us hope it will prove to be so
in 1896.' Letter dated 9 April 1896 from J.
Rattray & Son, Dunedin to James Preston,
Longlands Station, Maniototo, Otago:
Hocken Collections, MS-1272/105.
22 Cody family diaries, Lime Hills, northern
Southland: Hocken Collections, AG-736.
23 P. G. Holland, 'Plants and Lowland South
Canterbury Landscapes', *New Zealand
Geographer*, vol. 44, 1988, pp. 50–60.
24 The first of those statements is from
Minister of Lands, *The Albury Settlement,
South Canterbury*, Ministry of Lands,
Wellington, 1897, and the second is from
Minister of Lands, *The Kohika Settlement,
South Canterbury*, Ministry of Lands,
Wellington, 1901.
25 In the nineteenth century, a farmer's
advisor might have been a seed merchant
suggesting an appropriate mix of pasture
plants for the property, a nurseryman
suggesting varieties of fruit tree or shelter
plants for the property, or a neighbouring
farmer or manager suggesting a sustainable stocking density for the property.
An advisor as we understand that word – a
person trained to look at the entire farm
/ station operation and produce a sound
business plan – would not have been available much before the 1920s or 1930s.
26 Ida Valley Station letter books and diaries, Central Otago: Hocken Collections,
MS-0658.

27 J. Garner, *By His Own Merits: Sir John Hall
– Pioneer, Pastoralist and Pioneer*, Dryden
Press, Hororata, 1995.
28 Nathan and Varnham (eds), *The Amazing
World of James Hector*.
29 L. J. Wild, *The Life and Times of Sir James
Wilson of Bulls*, Whitcombe & Tombs,
Christchurch, 1953.
30 The authors of the following articles
stressed the importance of standardised
flocks of sheep (Anon., *New Zealand
Country Journal*, vol. 6, 1882, p.319)
and fruit (Anon., *New Zealand Country
Journal*, vol. 32, 1918, p. 319) to facilitate
exports to the northern hemisphere.
31 R. T. T. Forman, *Land Mosaics: The
Ecology of Landscapes and Regions*,
Cambridge University Press, New York,
1995; R. T. T. Forman and M. Godron,
Landscape Ecology, Wiley, New York,
1986; and C. D. Meurk and S. R. Swaffield,
'A Landscape Ecological Framework for
Indigenous Regeneration in Rural New
Zealand-Aotearoa', *Landscape and Urban
Planning*, vol. 50, 2000, pp. 129–44.
32 G. Byrnes, *Boundary Markers: Land
Surveying and the Colonisation of New
Zealand*, Bridget Williams Books,
Wellington, 2001.

Appendix: Words about Home

1 E. C. Richards (ed.), *Diary of E. R.
Chudleigh 1862–1921: Chatham Islands*,
Simpson & Williams, Christchurch,
1950; and E. R. Ward, *The Journal of
Edward Ward: 1850–51*, Pegasus Press,
Christchurch, 1951.
2 Ida Valley Station letter books and diaries, Central Otago: Hocken Collections,
MS-0658.
3 Bryce brothers' diaries (most entries were
by David Bryce), Lovells Flat, east Otago:
Hocken Collections, MS-0615.
4 James Preston's diaries, letter books and
personal papers, Southland, east Otago,
Maniototo, Waitaki Valley and Mackenzie
Country: Hocken Collections, MS-1271
and MS-1272.
5 The report of Jimmy Ford's lonely death
comes from the Longland Station diary
for 1899, entry dated 19 May (Hocken
Collections, MS-1271), and a copy of

6   Ramsay family diaries, Hyde, Central Otago: Hocken Collections, AG-680.

7   C. J. Ayton, *Charles J. Ayton, Diary 1899–1904: Goldminer, Rabbiter and Peat-Digger of Central Otago*, Maniototo Early Settlers Association, Naseby, 1982.

8   The following eight runs of diaries and letter books – arranged chronologically, as in Figure A.2 – proved especially informative. Rather than include a note for each reference to them, these sources will be clearly identified throughout the text by author and place, as well as the date of each citation. Full details follow, and are repeated in the list of references at the end of the book:
      Edward Chudleigh's diaries, mid-Canterbury: Richards (ed.), *Diary of E. R. Chudleigh 1862–1921*
      John Wither's diaries, Lake Wakatipu Basin, Central Otago: Hocken Collections, 89-149
      The Point Station diaries, mid-Canterbury: Canterbury Museum, Christchurch

James Preston's letter to F. W. Morrice, Public Trust Office, Dunedin, is in the station letter book (Hocken Collections, MS-1271).

Ida Valley Station letter books and diaries, Central Otago: Hocken Collections, MS-0658
Bryce brothers' diaries (most entries were by David Bryce), Lovells Flat, south Otago: Hocken Collections, MS-0615
Joseph Davidson's diaries, Switzers / Waikaia, northern Southland: Hocken Collections, AG-523
McMaster family diaries, letter books and cash ledgers, Tokarahi, Waitaki Valley, north Otago: Hocken Collections, MS-1011
James Preston's diaries, personal and business papers, Southland, east Otago, Maniototo, Waitaki Valley and Mackenzie Country: Hocken Collections, MS-1271 and MS-1272.

9   G. Wood and E. Pawson, 'Flows of Agricultural Information', in T. Brooking and E. Pawson (eds), *Seeds of Empire: The Environmental Transformation of New Zealand*, I. B. Taurus, London, 2011, pp. 139–59.

10   M. McKinnon (ed.), *Bateman New Zealand Historical Atlas*, Plate 54, 'Rural Society: People and Services in the 1880s and 1890s', David Bateman, Auckland, 1997.

BIBLIOGRAPHY

Unpublished farm and station diaries, letter books, correspondence and business records

Thomas and William Adams' diaries (selections transcribed by J. H. Beatie), Waihola, south Otago: Hocken Collections, MS-582.

John Barnicoat's diaries, Nelson and eastern South Island (transcript): Hocken Collections, MS-1451.

Benmore Station, upper Waitaki Valley, north Otago, manager's letter book: Hocken Collections, MS-3766.

Bryce brothers' diaries, (most entries by David Bryce), Lovells Flat, south Otago: Hocken Collections, MS-0615.

Cody family diaries, Lime Hills, northern Southland: Hocken Collections, AG-736.

Joseph Davidson's diaries, Switzers / Waikaia, northern Southland: Hocken Collections, AG-523.

Thomas Ferens' diaries, Waikouaiti, east Otago (transcript): Hocken Collections, MS-0440/016.

John Finlay's diaries, Makikihi, south Canterbury (transcript): Waimate Historical Society and Museum.

Grampian Hills Station diaries, Mackenzie Country (transcript): South Canterbury Museum, Timaru.

Joseph Greenwood's diaries, southern North Island, Wellington, Banks Peninsula and north Canterbury (transcript): Alexander Turnbull Library, Wellington, MS-PAPERS-4882.

John Gunn's diaries, Hook, south Canterbury: Hocken Collections, MS-1070.

John Hall's diaries, Edendale, Southland: Hocken Collections, MS-0292.

Ida Valley Station diaries and letter books, Central Otago: Hocken Collections, MS-0658.

Dickson Jardine's diaries, Otekaike, north Otago: Hocken Collections, AG-659

Kauru Hill diaries, north Otago: North Otago Museum, Oamaru, Box 59/33d.

Macdonald family diaries, Orari Station, south Canterbury: in private hands.

John Maddison's diaries, Southbridge-Leeston, mid-Canterbury: Canterbury Museum, Christchurch, ARC 1900.325.

McMaster family diaries, letter books and cash ledgers, Tokarahi, Waitaki Valley, north Otago: Hocken Collections, MS-1011.

James Menzies' diaries, Mataura, Southland: Hocken Collections (transcript), MS-3178/053.

James Miller's diaries, Outram-Maungatua, east Otago: Hocken Collections, 91-010.

Archibald Morton's letter book, Haldon Station, Mackenzie Country: Hocken Collections, MS-1272/064.

Mount Nicholas Station diaries, Lake Wakatipu basin (partial transcript): Hocken Collections, MS-1270-2-3/002.

Joseph Munnings' diaries, Christchurch and environs (transcript): Canterbury Museum, Christchurch, ARC 1990.50.

James Murison's diaries, Puketoi Station, Maniototo, eastern Otago (transcript): Hocken Collections, ARC-0359.

New Zealand and Australian Land Company, Levels Station, south Canterbury, manager's letter book: South Canterbury Museum, Timaru.

Francis Pillans' diaries, Inch Clutha, south Otago (transcript): Hocken Collections, MS-0616.

The Point diaries, mid-Canterbury: Canterbury Museum, Christchurch.

James Preston's diaries, personal and business papers, Southland, Maniototo, east Otago, Waitaki Valley and Mackenzie Country: Hocken Collections, MS-1271 and MS-1272.

James Preston's invoices and receipts relating to Centrewood Station, east Otago: Hocken Collections, MS-1272/090.

Raincliff Station diaries, south Canterbury (transcript): South Canterbury Museum, Timaru.

Ramsay family diaries, Hyde, on the fringes of Central Otago: Hocken Collections, AG-680.

Kate Sheath's diaries, Albury, south Canterbury (typed transcript): Waimate Historical Society and Museum.

William Shirres' letter book, Aviemore Station, Waitaki Valley, letter dated 20 June 1884: Hocken Collections, 291*MS-0635.

Frederick and William Teschemaker's diaries, Otaio Station, south Canterbury: South Canterbury Museum, Timaru.

Frederick Teschemaker's diaries, south Canterbury and east Otago (transcript): Hocken Collections, MS-0446.

Nugent Wade's letter books, Taipo Hill, Maheno, north Otago: South Canterbury Museum, Timaru.

John Wither's diaries, Lake Wakatipu basin, Central Otago (transcript): Hocken Collections, 89-149.

Books, book chapters, theses and published diaries

Acland, L. G. D., *The Early Canterbury Runs*, Whitcombe & Tombs, Christchurch, 1951.

Andersen, J. C., *Jubilee History of South Canterbury*, Whitcombe & Tombs, Auckland, 1916.

Anderson, K., *Predicting the Weather: Victorians and the Science of Meteorology*, University of Chicago Press, Chicago, 2005.

Anon., *Letters from Settlers and Labouring Emigrants in the New Zealand Company's Settlements of Wellington, Nelson, and New Plymouth, from February 1842 to January 1843*, Smith, Elder & Co., London, 1843.

Ayton, C. J., *Charles J. Ayton, Diary 1899–1904: Goldminer, Rabbiter and Peat-Digger of Central Otago*, Maniototo Early Settlers Association, Naseby, 1982.

Barker, Lady Mary, *Station Life in New Zealand*, Whitcombe & Tombs, Christchurch, 1950.

Bathgate, J., *New Zealand: Its Resources and Prospects*, W. & R. Chambers, London, 1880.

Bawden, P., *The Years Before Waitangi: A Story of Early Maori / European Contact in New Zealand*, published by the author, Auckland, 1987.

Beaglehole, J. C. (ed.), *The Endeavour Journal of Joseph Banks*, second edition, Public Library of New South Wales and Angus & Robertson, Sydney, 1963.

Beattie, J. H. (collated by A. Anderson), *Traditions and Legends of the South Island Maori: Collected from Natives of Murihiku (Southland, New Zealand)*, Cadsonbury Publications, Christchurch, 2004.

Best, E., *The Astronomical Knowledge of the Maori*, Monograph 3, Dominion Museum, Wellington, 1922.

Best, E., *The Maori As He Was: A Brief Account of Maori Life As It Was In Pre-European Days*, New Zealand Board of Science and Art, Manual 4, Dominion Museum, Wellington, 1934.

Broome, Lady Mary, *Colonial Memories*, Smith, Elder & Co., London, 1904.

Buchanan, J., *Manual of the Indigenous Grasses of New Zealand*, Colonial Museum, Wellington, 1880.

Burrows, C. J., *Julius Haast in the Southern Alps*, Canterbury University Press, Christchurch, 2005.

Byrnes, G., *Boundary Markers: Land Surveying and the Colonisation of New Zealand*, Bridget Williams Books, Wellington, 2001.

Clark, A. H., *The Invasion of New Zealand by People, Plants and Animals: The South Island*, Rutgers University Press, New Brunswick, 1949.

Clifton, N. C., *New Zealand Timbers: The Complete Guide to Exotic and Indigenous Woods*, GP Publications, Wellington, 1991.

Crawford, S. S., *Sheep and Sheepmen of Canterbury, 1850–1914*, Simpson & Williams, Christchurch, 1949.

Cresswell, D., *The Story of Cheviot*, Cheviot County Council, Hawarden, 1951.

Cruise, R. A., *Journal of a Ten Month's Residence in New Zealand*, Longman, Hurst, Rees, Orme, Brown & Green, London, 1823, reprinted by Pegasus Press, Christchurch, 1957.

Cumberland, K. B., *Soil Erosion in New Zealand: A Geographic Reconnaissance*, Whitcombe & Tombs
in collaboration with the Soil Conservation and Rivers Control Council, Wellington, 1947.

Davey, A. J., *Daybreak in Geraldine County, 1877–1952*, privately published, Geraldine, 1952.

de Bray, L., *Manual of Old-Fashioned Flowers*, Oxford Illustrated Press, Sparkford, 1984.

Dominy, M., *Calling the Station Home: Place and Identity in New Zealand's High Country*, Rowman &
Littlefield, Lanham, 2001.

Dueck, D., *Strabo of Amasia*, Routledge, London, 2000.

Edgar, E. and H. E. Connor, *Flora of New Zealand. Volume V, Graminae*, Manaaki Whenua Press,
Lincoln, 2000.

Evison, H. C., *Te Wai Pounamu the Greenstone Island: A History of the Southern Maori During the
European Colonisation of New Zealand*, Aoraki Press, Wellington, 1993.

Fearon, K. J., *Te Wharau*, Netherton Grange Publications, Masterton, 1993.

Forman, R. T. T., *Land Mosaics: The Ecology of Landscapes and Regions*, Cambridge University Press,
New York, 1995.

Forman, R. T. T. and M. Godron, *Landscape Ecology*, Wiley, New York, 1986.

Garner, J., *By His Own Merits: Sir John Hall – Pioneer, Pastoralist and Pioneer*, Dryden Press,
Hororata, 1995.

Gillespie, O. A., *South Canterbury: A Record of Settlement*, South Canterbury Centennial History
Committee, Timaru, 1958.

Godley, J. R. (ed.), *Letters from Early New Zealand by Charlotte Godley 1850–1853*, Whitcombe &
Tombs, Christchurch, 1951.

Gordon, M., *The Garden of Tane*, A. H. & A. W. Reed, Wellington, not dated.

Grant, S. and J. S. Foster, *New Zealand: A Report on its Agricultural Conditions and Prospects*,
G. Street, London, 1880.

Guthrie-Smith, H., *Tutira: The Story of a New Zealand Sheep Station*, Blackwood, Edinburgh,
1953.

Hector, J., *Handbook of New Zealand*, Lyon & Blair, Wellington, 1880.

Holland, P., K. O'Connor and A. Wearing, 'Remaking the Grasslands of the Open Country', in
E. Pawson and T. Brooking (eds), *Environmental Histories of New Zealand*, Oxford University
Press, Melbourne, 2002, pp. 69–83.

Holland, P., P. Star and V. Wood, 'Pioneer Grassland Farming: Pragmatism, Innovation and
Experimentation', in T. Brooking and E. Pawson (eds), *Seeds of Empire: The Environmental
Transformation of New Zealand*, I. B. Tauris, London, 2011, pp. 51–72.

Holland, P., J. Williams and V. Wood, 'Learning About the Environment in Early Colonial New
Zealand', in T. Brooking and E. Pawson (eds), *Seeds of Empire: The Environmental Transformation
of New Zealand*, I. B. Tauris, London, 2011, pp. 34–50.

'Hopeful', *'Taken in': Being a Sketch of New Zealand Life*, W. H. Allen, London, 1887.

Humboldt, A. von (translated by E. C. Otte), *Cosmos: A Sketch of a Physical Description of the
Universe*, five volumes, Harper, New York, 1870–1871.

Hursthouse, C., *New Zealand, the 'Britain of the South'; With a Chapter on the Native War, and our
Future Native Policy*, second edition, E. Stanford, London, 1861.

Jacobson, H. C. and J. W. Stack, *Tales of Banks Peninsula*, third edition, Akaroa Mail, Akaroa, 1917.

Liebig, J., *Letters on Modern Agriculture*, Walton & Maberly, London, 1859.

Lovelock, J., *Gaia: A New Look at Life on Earth*, Oxford University Press, Oxford, 1987.

Lovelock, J., *Gaia: The Practical Science of Planetary Medicine*, Allen & Unwin, Sydney, 1991.

Maling, P. B. (ed.), *The Torlesse Papers: The Journals and Letters of Charles Obins Torlesse
Concerning the Foundation of the Canterbury Settlement in New Zealand 1848–51*, Pegasus Press,
Christchurch, 1958.

McIntosh, A. D. (ed.), *Marlborough: A Provincial History*, Marlborough Provincial Historical
Committee, Blenheim, 1940.

McKinnon, M. (ed.), *Bateman New Zealand Historical Atlas*, David Bateman, Auckland, 1997.

McIntyre, W. D. (ed.), *The Journal of Henry Sewell, 1853–1857*, two volumes, Whitcoulls,
Christchurch, 1980.

McLean, W. H., *Rabbits Galore – on the Other Side of the Fence*, Reed, Wellington, 1966.

Morris, M. A., 'History of Christchurch Home Gardening from Colonisation to the Queen's Visit', PhD thesis, University of Canterbury, 2006.

Nathan, S. and M. Varnham (eds), *The Amazing World of James Hector*, Awa Science, Wellington, 2008.

Nicholson, M., *The Environmental Revolution: A Guide to the New Masters of the World,* Penguin Books, Harmondsworth, 1972.

Nicol, E. R. (compiler), *Common Names of Plants of New Zealand*, Maanaki Whenua Press, Lincoln, 1997.

Pawson, E. and P. Holland, 'People and the Land', in M. Winterbourn, G. Knox, C. Burrows and I. Marsden (eds), *The Natural History of Canterbury*, third edition, Canterbury University Press, Christchurch, 2008.

Peden, R. L., *Making Sheep Country: Mt Peel Station and the Transformation of the Tussock Lands*, Auckland University Press, Auckland, 2011.

Potts, T. H., *Out in the Open: A Budget of Scraps of Natural History Gathered in New Zealand*, Lyttelton Times Company, Christchurch, 1882.

Preston, F. L., *A Family of Woolgatherers*, John McIndoe, Dunedin, 1978.

Reeves, W. P., *New Zealand, and Other Poems*, Grant Richards, London, 1898.

Reeves, W. P., *New Zealand*, Adam & Charles Black, London, 1908.

Relph, E. C., *Place and Placelessness*, Pion, London, 1976.

R. G. S., 'Seed Mixtures: Controversies Past, Present and Future', in I. H. Hunter (ed.), *Bailliere's Encyclopaedia of Scientific Agriculture*, Bailliere, London, 1931, pp. 1104–18.

Richards, E. C. (ed.), *Diary of E. R. Chudleigh 1862–1921: Chatham Islands*, Simpson & Williams, Christchurch, 1950.

Selfe, H. S., *Canterbury New Zealand in 1862*, G. Street, 'New Zealand Examiner' Office, London, 1862.

Semple, E. C., *Influences of the Geographical Environment, on the Basis of Ratzel's System of Anthropo-Geography*, H. Holt, New York, 1911.

Sharp, C. A. (ed.), *The Dillon Letters: The Letters of the Hon. Constantine Dillon, 1842–1853*, A. H. & A. W. Reed, Wellington, 1954.

Shortland, E., *The Southern Districts of New Zealand; A Journal with Passing Notices of the Customs of the Aborigines*, Longman, Brown, Green & Longman, London, 1851.

Siegel, S., *Nonparametric Statistics for the Behavioural Sciences*, McGraw-Hill, New York, 1956.

Speight, R. A. Wall and R. M. Laing (eds), *Natural History of Canterbury*, Simpson & Williams, Christchurch, 1927.

Stack, J. W., *Through Canterbury and Otago with Bishop Harper in 1858–60*, Akaroa Mail, Akaroa, not dated.

Stack, J. W., *Kaiapohia: The Story of a Siege*, Whitcombe & Tombs, Christchurch, 1893.

Stack, J. W., edited by A. H. Reed, *A White Boy Among the Maoris in the 'Forties': Pages from an Unpublished Autobiography of James West Stack*, A. H. Reed, Dunedin, 1934.

Studholme, E. M. L., *Reminiscences of 1860*, Whitcombe & Tombs, Christchurch, not dated.

Tansley, A. G., *Introduction to Plant Ecology*, G. Allen, London, 1946.

Tothill, G. C., *Laying Down Land in Permanent Grasses*, Southland Times Company, Invercargill, 1893.

Tuan, Yi-fu, *Space and Place: The Perspective of Experience*, University of Minnesota Press, Minneapolis, 1977.

Vincent, G. T., 'Sports and Other Signs of Civilisation in Colonial Canterbury, 1850–1890', PhD thesis, University of Canterbury, 2002.

Wall, A., 'A Sheep Station Glossary', in L. G. D. Acland, *The Early Canterbury Runs*, Whitcombe & Tombs, Christchurch, 1951, pp. 355–411.

Ward, E. R., *The Journal of Edward Ward: 1850–51*, Pegasus Press, Christchurch, 1951.

Ward, J., *Information Relative to New Zealand, Compiled for the Use of Colonists*, second edition, John W. Parker, London, 1840.

Wild, L. J., *The Life and Times of Sir James Wilson of Bulls*, Whitcombe & Tombs, Christchurch.

Williams, J., 'E Pākihi Hakinga a Kai', PhD thesis, University of Otago, 2004.

239

Wodzicki, K. A., *Introduced Mammals of New Zealand: An Ecological and Economic Survey*, Bulletin No. 98, Department of Scientific and Industrial Research, Wellington, 1950.

Wood, G. and E. Pawson, 'Flows of Agricultural Information', in T. Brooking and E. Pawson (eds), *Seeds of Empire: The Environmental Transformation of New Zealand*, I. B. Taurus, London, 2011, pp. 139–59.

Woodhouse, A. E., *George Rhodes of the Levels and his Brothers*, Whitcombe & Tombs, Auckland, 1937.

Journal articles

'Agricola', 'Hedge Plants', *New Zealand Country Journal*, vol. 1, 1877, pp. 94–95.

Anon., 'The First Pastoral and Agricultural Show, *New Zealand Country Journal*, vol. 1, 1877, pp. 21–26.

Anon., *New Zealand Country Journal*, vol. 6, 1882, p. 319

Anon., 'Judging at Shows', *New Zealand Country Journal*, vol. 13, 1889, pp. 456–58.

Anon., 'Mixtures of Grass Seed', *New Zealand Stock and Station Journal*, June 1902, p. 269.

Anon., 'The Purity of Seed', *New Zealand Farmer*, vol. 25, 1904, pp. 964–65.

Anon., 'Permanent Pastures', *New Zealand Farmer, Stock and Station Journal*, vol. 27, 1906, pp. 713–15.

Anon., 'The Farm; Putting Down to Grass', *New Zealand Farmer, Stock and Station Journal*, vol. 34, 1913, pp. 37–38.

Anon., *New Zealand Country Journal*, vol. 32, 1918, p. 319.

Armstrong, J. B., 'The Native Grasses', *New Zealand Country Journal*, vol. 3, 1879, pp. 201–4.

Armstrong, J. F., 'On the Naturalized Plants of the Province of Canterbury', *Transactions of the New Zealand Institute*, vol. 4, 1871, pp. 284–90.

Barrows, H. H., 'Geography as Human Ecology', *Annals of the Association of American Geographers*, 1923, vol. 13, pp. 1–14.

Challenger, S., 'Pioneer Nurserymen of Canterbury, New Zealand (1860–65)', *Garden History*, vol. 7, 1979, pp. 25–64.

Cockayne, A. H., 'Impure Seed: The Source of our Weed Problem', *New Zealand Journal of Agriculture*, vol. 4, 1912, pp. 437–43.

Cockayne, A. H., 'The Surface-Sown Grass Lands of New Zealand', *New Zealand Journal of Agriculture*, vol. 7, 1913, pp. 465–75.

Cockayne, A. H., 'Seed-Impurities', *New Zealand Journal of Agriculture*, vol. 13, 1916, pp. 208–12.

Cockayne, A. H., 'Surface-Sown Grass Mixtures: Some Recent Changes', *New Zealand Journal of Agriculture*, vol. 14, 1917, pp. 169–77.

Cockayne, A. H., 'The Grasslands of New Zealand', *New Zealand Journal of Agriculture*, vol. 17, 1918, pp. 140–42.

Cockayne, A. H., 'The Grasslands of New Zealand: Component Species', *New Zealand Journal of Agriculture*, vol. 17, 1918, pp. 210–20.

Cockayne, L., 'An Economic Investigation of the Montane Tussock-Grassland of New Zealand. No. I. – Introduction', *New Zealand Journal of Agriculture*, vol. 18, 1919, pp. 1–9.

Crawford, J. C., 'On Hedges and Hedge Plants', *Transactions of the New Zealand Institute*, vol. 9, 1876, pp. 203–6.

Cronon, W., 'The Trouble With Wilderness: or, Getting Back to the Wrong Nature', *Environmental History*, vol. 1, 1995, pp. 7–55.

Cumberland, K. B., 'A Century's Change: Natural to Cultural Vegetation in New Zealand', *Geographical Review*, vol. 31, 1941, pp. 529–54.

Dixon, M., 'Permanent Pasture Grasses', *New Zealand Country Journal*, vol. 10, 1886, pp. 35–38.

Firth, J. C., 'On Forest Culture', *New Zealand Country Journal*, vol. 3, 1879, pp. 142–49.

Garcia-Herrera, R., H. F. Diaz, R. R. Garcia, M. R. Prieto, D. Barriopedro, R. Moyano and E. Hernandez, 'A Chronology of El Niño Events from Primary Documentary Sources in Northern Peru', *Journal of Climate*, vol. 21, 2008, pp. 1948–62.

Gray, E., 'The Early History of Ryegrass Seed Production in Scotland', *Seed Trade Review*, April 1952, pp. 103–10.

Green, A. E., 'Native Trees and Shrubs as Hedge Plants', *New Zealand Journal of Agriculture*, vol. 4, 1912, pp. 444–48.

Guthrie-Smith, H., 'Grassland Evolution at Tutira', *New Zealand Journal of Agriculture*, vol. 40, 1930, pp. 84–92.

Holland, P. G., 'Plants and Lowland South Canterbury Landscapes', *New Zealand Geographer*, vol. 44, 1988, pp. 50–60.

Holland, P. G., 'Cultural Landscapes as Biogeographical Experiments: A New Zealand Perspective', *Journal of Biogeography*, vol. 27, 2001, pp. 39–43.

Holland, P. and P. Dixon, 'New Plants for the "Howling Wilderness"', *Bulletin of the Dunedin Rhododendron Group*, no. 34, 2006, pp. 18–29.

Holland, P., P. Dixon and V. Wood, 'Learning About the Weather in Early Colonial New Zealand', *Weather and Climate*, vol. 29, 2009, pp. 3–21.

Holland, P. and B. Mooney, 'Wind and Water: Environmental Learning in Early Colonial New Zealand', *New Zealand Geographer*, vol. 62, 2006, pp. 39–49.

Holland, P. and A. Wearing, 'By Choice: Plants Grown by Settlers in Lowland Canterbury', in M. Roche, M. McKenna and P. Hesp (eds), *Proceedings of the Twentieth New Zealand Geographical Society Conference*, New Zealand Geographical Society, Hamilton, 1999, pp. 50–54.

Holland, P. and V. Wood, 'Decorative Plants and Horticultural Exhibitions in Early New Zealand', *Bulletin of the Dunedin Rhododendron Group*, no. 36, 2008, pp. 15–28.

Hopkins, J. M., 'Native Bush and Excellent Stock Shelter: its Protection and Regeneration Well Worth While', *New Zealand Journal of Agriculture*, vol. 75, 1975, pp. 257–59.

Hutton, F. W., 'Cross-Fertilisation of Plants and its Relation to Agriculture', *New Zealand Country Journal*, vol. 7, 1883, pp. 331–34.

Johnson, R. J., 'There's a Place for Us', *New Zealand Geographer*, vol. 44, 1988, pp. 8–13.

King, D. N. T., J. Goff and A. Skinner, 'Maori Environmental Knowledge and Natural Hazards in Aotearoa-New Zealand', *Journal of the Royal Society of New Zealand*, vol. 37, 2007, pp. 59–73.

Kjægaard, T., 'A Plant that Changed the World: The Rise and Fall of Clover, 1000–2000', *Landscape Research*, vol. 28, 2003, pp. 41–49.

Levy, E. B., 'The Grasslands of New Zealand: Principles of Pasture Establishment', *New Zealand Journal of Agriculture*, vol. 23, 1921(a), pp. 259–65.

Levy, E. B., 'The Grasslands of New Zealand: Principles of Pasture Establishment', *New Zealand Journal of Agriculture*, vol. 23, 1921(b), pp. 321–30.

Levy, E. B., 'The Grasslands of New Zealand: Preliminary Ecological Classification of Species', *New Zealand Journal of Agriculture*, vol. 30, 1925, pp. 357–74.

Levy, E. B., 'Observations Relative to Pasture Seed-Mixtures', *Proceedings of the Fifth Conference of the New Zealand Grasslands Association*, 1936, pp. 33–40.

McConnell, P., 'Cultivation', *Journal of the Department of Agriculture*, 15 December 1911, pp. 448–53.

Meurk, C. D. and S. R. Swaffield, 'A Landscape Ecological Framework for Indigenous Regeneration in Rural New Zealand-Aotearoa', *Landscape and Urban Planning*, vol. 50, 2000, pp. 129–44.

'Ovis', 'Permanent Pastures and Alternative Farming', *New Zealand Country Journal*, vol. 13, 1889, pp. 451–56.

'Ovis', 'Pastures', *New Zealand Country Journal*, vol. 21, 1897, pp. 393–97.

Parker, O. G., 'Laying Down Land to Grass and its Subsequent Treatment', *New Zealand Country Journal*, vol. 5, 1881, pp. 401–4.

Patullo, P., 'Farming Experiments', *New Zealand Country Journal*, vol. 22, 1898, pp. 374–79.

Pawson, E. and P. Holland, 'Lowland Canterbury Landscapes in the Making', *New Zealand Geographer*, vol. 61, 2005, pp. 167–75.

Potts, H. W., 'The Exhaustion of our Grazing Lands', *New Zealand Country Journal*, vol. 21, 1897, pp. 507–9.

Potts, T. H., 'The Kea – Progress of Development', *Nature*, vol. 4, 1871, p. 489.

Potts, T. H., 'Change of Habits in Animals and Plants', *Nature*, vol. 5, 1872, p. 262.

Sears, P. D., 'Exploitation of High Production Pastures in New Zealand', *Proceedings of the Ninth Conference of the New Zealand Ecological Society*, 1962.

Sewell, T. G., 'Improvement of Tussock Grassland', *New Zealand Journal of Agriculture*, vol. 81, 1950, pp. 503–17.

Siebert, C., 'Food Ark', *National Geographic*, vol. 221, 2011, pp. 108–31.

Simmonds, J. H., 'Wind-Breaks for Fruit Farms', *New Zealand Journal of Agriculture*, vol. 15, 1917, pp. 253–62.

Simson, A., 'Pasturage', *New Zealand Country Journal*, vol. 10, 1886, pp. 291–95.

Smith, J. M., 'Some Aspects of the Extreme Simplification of Pasture Seed-Mixtures', *Proceedings of the Fifth Conference New Zealand Grasslands Association*, 1936, pp. 41–45.

Taylor, W. H., 'Native Plants: the Work of Collection and Establishment', *New Zealand Journal of Agriculture*, vol. 7, 1913, pp. 61–64.

Wilkin, R., 'Grasses and Forage Plants Best Adapted to New Zealand', *New Zealand Country Journal*, vol. 1, 1877, pp. 3–12.

Wilkin, R., 'A Gossip About Grasses', *New Zealand Country Journal*, vol. 3, 1879, pp. 294–98.

Wilkin, R., 'Does the Continuous Grazing of Land with Sheep Cause it to Deteriorate in Value?', *New Zealand Country Journal*, vol. 5, 1881, pp. 3–4.

Wilson, B., 'Stock to Fit the Environment', *New Zealand Farmer*, August 1982, pp. 184–99 and 201–7.

Wilson, J. G., 'Grasses', *New Zealand Farmer, Bee and Poultry Journal*, vol. 20, 1900, pp. 457–58.

Wilson, J. G., 'Grasses', *New Zealand Farmer, Bee and Poultry Journal*, vol. 21, 1901, pp. 1–3.

Yeates, J. S., 'Farm Trees and Hedges', *Massey Agricultural College Bulletin*, number 12, 1948.

Official documents

Anon., *New Zealand Official Yearbook*, Government Printer, Wellington, 1894, p. 276.

Appendix to the Journals of the House of Representatives of New Zealand, F-9, 1870; I-5, 1886; C-17, 1887; H-11, 1887; H-18, 1888; H-31, 1888.

Evans, B. L., 'Grassland Research in New Zealand', in *New Zealand Official Year-Book*, Government Printer, Wellington, 1960, pp. 1234–64.

Minister of Lands, *The Albury Settlement, South Canterbury*, Ministry of Lands, Wellington, 1897.

Minister of Lands, *The Kohika Settlement, South Canterbury*, Ministry of Lands, Wellington, 1901.

Ministry for the Environment, *Environment New Zealand 2007*, Ministry for the Environment, Wellington, 2007.

Miscellaneous documents

J. C. Andersen, field book containing notes about Māori nomenclature and legends: Alexander Turnbull Library, Wellington, MS-PAPERS-0148-112.

J. Herries Beattie, handwritten notes, 1915–1922: Hocken Collections, MS-0181.

J. Herries Beattie, 'Record of Interviews with Maori in Canterbury – Section XXVIII', handwritten notes: Hocken Collections, MS-0181/005.

Clutha County, outward letter book 1879–81, Peter Nelson, County Clerk, to J. M. McKenzie, Member of the House of Representatives, letter dated 1 May 1889: Hocken Collections, 323*AG-253-001/001.

J. Cowan, loose-leaf folder containing miscellaneous items: Alexander Turnbull Library, Wellington, MS-PAPERS-0039.

Manchester and Goldsmith, ledger books for Te Waimate Station, south Canterbury: Waimate Historical Society and Museum.

New Zealand Loan & Mercantile Agency, outwards correspondence from the Gore to Invercargill branches: Hocken Collection, AG-875/006.

James Preston to the Chairman of the Land Board, Christchurch, applying for relief under The Pastoral Tenants Act 1895: Hocken Collections, MS-1271/016.

James Preston to William Grant, letter dated 22 November 1904: Hocken Collections, MS-1272/066.

J. Rattray & Son, Dunedin to J. H. Preston, Longlands Station: Hocken Collections, MS-1272/105.

New Zealand newspapers

Akaroa Mail
Australasian
Bruce Herald
Canterbury Times
The Colonist
Dunstan Times
Grey River Argus
Hawke's Bay Herald
Lake Wakatip Mail
Lyttelton Times
Mount Ida Chronicle
Nelson Examiner

Oamaru Times
Otago Daily Times
Otago Witness
The Press
Rangitikei Advocate
Southern Cross
Southland News
Southland Times
Timaru Herald
Weekly Press
The Weekly Times
West Coast Times

INDEX

acclimatisation societies, 12

Acland family, 105, 218

'Agricola', 124

agricultural and pastoral associations and shows, 172, 178–83, 184, 185–6, 187, 200; *see also* Agricultural and Pastoral Society of Otago; Canterbury Agricultural and Pastoral Association; Clutha and Matua Agricultural and Pastoral Association; Nelson Agricultural Association; Northern Agricultural and Pastoral Association; Pahiatua Agricultural and Pastoral Association; Timaru Agricultural and Pastoral Association

Agricultural and Pastoral Society of Otago, 180

agronomy, 5, 8, 13, 107, 112, 121, 199, 208, 209

Akaroa, 22, 99, 123, 176

Akaroa Mail, 117

Albury Settlement, 204

Alport, H. E., 175

American Clydesdale Association, 182

American Live Stock Journal, 182

Andersen, Johannes, 25–26, 29

animal husbandry, 107, 110

animals, introduced, 3, 4, 12, 96, 105, 107–8, 109, 139, 140, 146, 155, 192, 193, 197, 199, 210

animals, native, 13, 17, 30–31, 88, 89, 90, 91–92, 94–95, 96–97, 100, 103, 105, 108, 126, 136, 171, 193, 194, 200, 208, 210–11; *see also* birds; Canterbury, native plants and animals in; fish, native; pest animals, native species as

Amuri district, 146

Aoraki, *see* Mount Cook

Arahura, 22–23

Armstrong, J. B., 103, 104, 105, 208, 210

Armstrong, J. F., 103, 208, 210

Arowhenua, 28, 33

Arrowtown, 48, 50, 141, 218

Arthur's Pass, 23, 134

Ashburton, 179

Ashley River, 72

Athenaeum, 8, 10

Australasian, 143

Australia, 48, 75, 105, 188, 200, 208, 213; fences in, 125; newspapers from, 125, 143, 178, 212; pasture grasses in, 119; plants from, 124, 126; timber from, 103; weather in, 48, 75, 77, 80; *see also Australasian*; *Hobart Town Mercury*; *Melbourne Age*; *Monaro Mercury*; weather patterns, comparisons with Australia; *Yeoman*

Aviemore Station, 64, 65, 78, 83, 101, 148–50, 152, 159–60

Avon River, 72, 74

Ayton, Charles, 218

bad weather, effects of on farming, 37, 52–56, 62, 66, 67, 69–70, 73–74, 81, 82–84, 85, 87, 189, 198–9

Baily's Magazine, 183

Balclutha, 75, 97, 146, 164, 184, 185, 186, 213, 219

Banks Peninsula, 5, 22, 26, 34, 47, 72, 89, 99, 106, 123, 140–1, 177

Banks, Joseph, 193

Barker, Lady Mary, 27, 30–31

Barnicoat, John, 10, 25, 44, 47

barometers, *see* weather forecasting and record-keeping

Barrows, Harlan, 191

Bawden, P., 18–19, 32

Bealey River, 23

Beattie, Herries, 18

Begg, John, 185

Bell and Bradford, 9

Bell Report, 19

Ben Ohau Station, 76, 83, 159–60

Bengal Advertiser, 8

Best, Elsdon, 17, 18, 31

Birch Hill Station, 21

birds: forest and wetland, 13; native, 13, 26, 30, 93, 94, 99–100; of prey, 148; seed-eating, 15, 165; *see also* Māori: botanical and zoological knowledge of; pets, native birds as

Black Forest Station, 148

Blenheim, 22, 72

Blueskin Bay, 29, 141

bone dust, use of, 115, 143, 144

books, 5, 8, 9–10, 27, 41–42, 102, 105, 121, 175, 186, 188, 191

Bowen, Sir George, 63–64